HISTORIC SHIPWRECKS AND RESCUES ON LAKE MICHIGAN

MICHAEL PASSWATER

Published by The History Press
Charleston, SC
www.historypress.com

First published 2022

Manufactured in the United States

ISBN 9781467151962

Library of Congress Control Number: 2022939460

Contents

Preface

There has always been fascination with history and the "whys" for things that happen. Some of the whys are obvious. Case in point: the *Henry B. Smith* was lost on Lake Superior in the great storm of 1913. She went out into the gale from the harbor at Marquette Michigan, with all her hatches open, and wasn't seen again for one hundred years. Shipwreck hunters found her in deep water north of Marquette in 2013. The reason she was lost is pretty obvious, but some of the "whys" are not so obvious. For instance, why did her captain take her out in such a bad storm with all her hatches open?

It was one of those questions that led to this book. It started out very simple: Why was Napier Avenue in St. Joseph, Michigan, named that way it was? Was the avenue named after Nelson Napier, the captain of the ill-fated *Alpena*? The answer was yes, but that led to an article about the Coast Guard naming one of its new cutters *Joseph Napier*. This led to the questions of why did they name this cutter after him and was he related to Nelson? After more research, of course, more questions came up. Joseph Napier was the first keeper of the lifesaving station in St. Joseph, and he was indeed related to Nelson. He also was the first man in the lifesaving service to receive a gold medal for his heroics in performing a rescue. He had a few other amazing rescues to his credit as well, before there was a lifesaving service and while he was the harbormaster in Chicago. This book got its start because of one simple question: Why was Joseph Napier important enough to get a ship named after him?

The fascinating history of the U.S. Life-Saving Service is amazing. The men of the early service were the local heroes in each town that had a station. This was before the lightbulb was developed and before the general public started envying the superficial heroes from sport teams and movies. The local population watched these men train every day, knowing that at any moment, they could be called into action to risk their lives to be able to save the lives of shipwrecked crews. Their job was to go out into a raging storm and rescue the lives of sailors in distress. The history of the U.S. Life-Saving Service has been documented by many authors, like Fred Stonehouse. The published stories of the daring storm warriors in numerous books are enough to take a person's breath away. This service was started out on the East Coast of the fledgling United States by volunteer organizations. It became apparent, due to an increase in shipwrecks and loss of life, that more organization, more maintenance of equipment and more training of its personnel was needed to save lives. The fledgling organization eventually became a department of the federal government.

As our country grew and industrial commerce increased, the number of ships needed to transport those goods had to increase commensurately. Of course, with the increase in shipping, the number of shipwrecks and loss of life increased. This created a huge outcry for the government to act and reduce the loss of life from shipwrecks. As the country grew and spread to the West, the lifesaving service made its way to the Great Lakes.

There are many stories about the lifesaving service on the Great Lakes—stories about astonishing rescues, mostly on Lakes Superior and Huron. Few of those stories say anything about Lake Michigan, which was one of the most traveled of the Great Lakes because of the big cities of Chicago and Milwaukee and the vast supply of timber along the shores of Michigan and Wisconsin. As this country grew, the easiest way to transport material or travel was by water. Interstate highways weren't even imagined at the time, and the roads that did exist were crude and underdeveloped by today's standards. To transport anything by wagon took a lot of effort. As cities like Chicago and Milwaukee grew, the demand for produce, wood and other raw materials also increased. This caused cities on Lake Michigan—like Saint Joseph, New Buffalo, Michigan City, South Haven and Holland and all the cities north of them—to grow as well. The farmers and lumbermen used these ports to ship their products to market. Most of these cities were located on a river with a channel out to Lake Michigan, with some sort of breakwater to protect the entrance. A schooner running under sail from a storm and coming into these ports was akin to trying to thread a needle. It was difficult and risky at best, and most of the time, the ship would end up missing the entrance and hit a sandbar offshore, where the wind and the waves would turn the ship into a wreck. As the mode of power shifted from sails to steam, this situation improved a little, but it was still a risky proposition to enter these ports during a storm.

Sandbars were created by the rivers flowing out of the channel into the lake. These sandbars always posed a danger by dynamically changing due to the whims of the winds acting on the waves. It was constant challenge to a ship running into a harbor. The ship would rise and fall with high waves, and with the sandbars in the lake changing, they could easily damage themselves in an area that was clear the day before. In many a storm, the ship would strike a sandbar, disabling the rudder or propeller and rendering it useless to control. It was during these times of peril that the lifesaving service would be called to action to save the lives of the sailors and passengers on these ships.

Living close to the east coast of Lake Michigan, I felt that these men's stories needed to be told. Hopefully you will find the following stories about these brave men enjoyable and come to appreciate the risks that they took in service to others.

Acknowledgements

Wes Oleszewski, a Michigan native who wrote many historical narratives, wisely said that even though an author gets his name on a book cover, the book is really a team effort made by the contributions of many people. I was always struck by that humble statement and did not realize how true it was until I wrote this book.

My first acknowledgment goes to my wife, Deb, even though this was written nine years after she became an angel. When we first met, she insisted that I love not only her but also the Great Lakes. Her family always vacationed in the Upper Peninsula of Michigan, and if we were going to have a relationship, I would need to love that area too. So, one extended weekend, with Deb, my girlfriend at the time, and her brother and sister and their better halves, we all went to Mackinac and spent the weekend exploring the area. By the end of that weekend, I knew two things: I loved my future wife *and* the Great Lakes and their history. Without her faith and encouragement in me, this book would not exist.

The people who work in the local libraries along the east coast of Lake Michigan also deserve my thanks. Their help, especially during the COVID-19 pandemic, was invaluable. These librarians include Jeanette Weiden at the Loutit District Library in Grand Haven; Clarie Gillespie and Elizabeth Appleton at the Maud Preston Library in Saint Joseph, Michigan; Brenda Norris at the Michigan City Library in Michigan City, Indiana; and Joseph Zappacosta of the Hackley Library in Muskegon, Michigan. I know there are others whose names I have forgotten, but I definitely appreciate their support.

I also credit other authors, including Fred Stonehouse, Wes Oleszewski, Dwight Boyer, Chris Kohl, Hugh Bishop, Jim Marshall, Dana Thomas Bowen and V.O. Vanheest, whose books I have spent countless hours reading. I especially want to include William Ratigan for his book *Great Lakes Shipwrecks and Survivals*. It was the first book on this subject I read, and I have been reading voraciously ever since.

I would also like to thank my friends in Duluth, Minnesota, whom I met at the Gales of November conference, as well as Fred Stonehouse again. Fred has spoken at that conference countless times, and I have attended hours of his lectures. I also had the pleasure of sharing drinks with him and his charming wife. Also, I thank my friends Cindi Baily and Virginia Stone—we have shared many stories. I also want to include my diving buddy and good friend Kent Ramsey. He has endured Chicago traffic on many occasions on the long drive up to Duluth with me and has spent a month's worth of time underwater exploring shipwrecks of the Great Lakes. I've come to trust this man with my life—he is a great diver and a great person. I thank all of my other dance and boating friends, who had to endure countless hours of my talking about shipwrecks. Also, thanks to my editor, Melinda Ruben of GreenSeed Consulting, for her skill and patience in correcting the grammar and wording in this book. Finally, I thank my mom, dad, brothers, sisters, sisters-in-law and brothers-in-law for their encouragement.

This book would not exist if it weren't for these special people.

What's in a Name?

On January 29, 2016, the fast Coast Guard cutter *Joseph Napier*, WPC-1115, was commissioned in San Juan, Puerto Rico. She was the fifteenth ship of the fast-response class of cutters that the Coast Guard had commissioned. The fast-response cutters have graceful and elegant lines. Underway, she is beauty in motion, but her true purpose was to respond quickly and efficiently to protect the United States and enforce its laws.

This new class of Coast Guard cutters is the latest in technology. At 154 feet and 353 tons, she can go through the water at twenty-eight knots. Her two main engines produce 5,766 horsepower each, and she can cruise 2,950 nautical miles. The cutters pack a MK 38-25mm cannon and four .50-caliber machine guns for close support. With a stern launch ramp, she can efficiently launch and recover a 26-foot fast cutter boat while underway. The cutters' main missions are coastal law enforcement, defense operations, illegal drug interception, illegal immigrant interception and interdiction. These cutters epitomize a standard of technological design that is highly efficient and stable.

So, what did it take for a person to have such an impressive class of ship named after him? Who exactly is Joseph Napier, and what did he do to deserve such an honor from the Coast Guard? To receive such an honor, a person must perform remarkable acts of courage and sacrifice in the line of duty. The distinction is given to those who put others above themselves and are willing to make the sacrifices necessary to prove the Coast Guard's motto, *Semper paratus*, "Always Ready."

Joseph Napier was born in Ohio in 1826 and had three brothers and five sisters. One brother, Nelson, was the captain of the ill-fated *Alpena* and was lost in a terrific storm in October 1880. His brother Andrew "Jack" died tragically on July 4, 1859, in St. Joseph, Michigan, when he lit a six-pounder cannon for the Independence Day celebration that exploded into dozens of fragments. One of those fragments struck and killed Jack. His other siblings were Orris, Amelia, Adeline, Olive, Nancy and Harriet.

In 1852, at the young age of twenty-six, Joseph became the harbor master in Chicago, Illinois, one of the busiest harbors on the Great Lakes. After a few years in Chicago, events transpired that would define him as the kind of person who deserves to be honored by the City of Chicago and the Coast Guard.

When the schooner *Merchant* approached Chicago in a horrifically turbulent storm on April 27, 1854, the waves and wind produced such dangerous conditions outside the harbor that they prohibited the captain of the schooner from entering. The raging maelstrom also prevented the Chicago Harbor tugs from assisting the *Merchant* in making safe passage into the harbor. So, the *Merchant*'s captain carefully chose a place off the Chicago breakwater and hoped that he could ride out the storm securely anchored to the bottom of Lake Michigan. Shortly after the anchor was dropped, the wind intensity increased, causing such chaotic conditions between the waves and wind that the frenzied combination rolled the *Merchant* onto her side. Four of her crew went over the side and were drowned in the violent waves. The three remaining crewmen clung to her exposed hull with the tenacity of men who just witnessed their shipmates die horrible deaths. Drenched by the icy waves washing over the hull, they could only hold tight to the hull and wonder how long it would take for the cold to sap their energy, leaving them to join their unfortunate shipmates.

Seeing the drama unfolding outside the harbor, Captain Joseph Napier immediately got eight men to volunteer to attempt to row a boat out to the distressed schooner. This was the only way to save these men and keep them from certain death. After a long, hard pull at the oars, Joseph and his volunteer crew were able to bring the three remaining sailors back to the safety of the harbor. For such heroic actions, the eight volunteers and Captain Joseph were awarded gold watches from the citizens of Chicago. Each watch was engraved with the following inscription: "For his noble and gallant efforts to rescue the crew of the Schooner Merchant, while in distress off the Port of Chicago, April 27, 1854."

The Men Who Received the Engraved Watches

Captain S.M. Johnson	Captain G.P. Ozier
Captain Joseph Napier	Captain Chas. J. Magill
Captain E.F. Drummong	Captain John Wiley
Mr. Joseph Nicholson	Mr. W.W. Niles
Mr. Quisling	

In September 1855, the sailing brig *Tuscarora*, during another violent storm on Lake Michigan, developed a leak in her hull, allowing large amounts of water to overcome her crew's desperate attempts to keep her afloat. Sitting low in the water, she hit a sandbar outside of the Chicago Harbor with enough force that she split wide open and started to break up. With no other options, the crew climbed the rigging to get out of the waves that were completely washing over her hull. This course, however, left them exposed to the freezing wind.

When the news of the *Tuscarora*'s condition was reported to Harbor Master Napier, he raised two volunteer crews to take two government lifeboats out to rescue the crew. Napier took command of one of the boats, while his brother Captain Andrew "Jack" commanded the other lifeboat. Captain Jack happened to be in the harbor on that stormy day with the propeller *Rossitter*. The men in the boats were as follows:

First Government Lifeboat

In Command:

Captain J.A. Napier	Harbor Master

Crew:

Captain Warren	Marine Inspector
Captain Rummage	Schooner: *Harvest*
Morris Evans	Seaman
Dennis Simmons	Seaman
George Golding	Seaman
John McElliott	Seaman

Second Government Lifeboat

In Command:

Captain A.J. Napier	Propeller: *Rossitter*

Crew:

Captain H.A. Gadsden	Brig: *Black Hawk*
Captain Jeffords	Brig: *Globe*
Captain Hiram Blood	Schooner: *Chapman*
Captain C.P. Morey	Schooner: *Lookout*
Captain C. Reed	Schooner: *Merdian*
Captain P.J. Mahoney	Schooner: *Maine*

As the news of the impending shipwreck spread throughout the city of Chicago, people braved the brutal weather to line the shore and watch the dramatic outcome of the rescue. As each boat left the protection of the Chicago River to head out into the turbulent lake, the waves were cresting five feet over each boat. Each lifeboat was skillfully guided by a Napier at the steering oar. As the lifeboats approached the *Tuscarora*, a cheer went up from the eleven crewmen still on board and clinging to life in her rigging. The lifeboats had to approach the *Tuscarora* from the lee side of the ship, where the wind and waves were slightly less intense but still large enough to prevent the boats from securing to the shipwreck. As the lifeboats reached as close as possible to the stricken ship, each crewman from the *Tuscarora* climbed down the rigging onto a storm-lashed deck and then literally jumped for his life into a lifeboat. The timing for each jump was extremely critical—it has to be at the exact moment when the boat and ship were at the same level. Any hesitation would cause the crewman to be crushed to death between the boat and ship or miss completely and suffer the ghastly death of drowning. The spectators on shore, with their eyes glued to the thrilling rescue, held their breath each time a crewman jumped. Shouts of joy erupted from the spectators after each crewman safely transitioned from the shipwreck into a lifeboat. After ten of eleven crewmen successfully transferred into the lifeboats, only her captain remained on board. Worried about surrendering her to the wreckers and losing his substantial investment in the ship, he declared that he would stay with the *Tuscarora* and share her fate. After several heart-stopping minutes in the rough lake and passionate pleas from his crew, he was finally persuaded that financial loss was better than the loss of his life. The captain made the daring leap and successfully landed into a boat.

With all the crew of the *Tuscarora* in lifeboats, the challenge ahead was the herculean task of getting everyone safely back on shore. Heading into the wind and waves, both Joseph and Jack were able to skillfully steer their boats through the mountainous waves and back into the protected waters of the

Chicago River. The brave rescuers and crew of the *Tuscarora* were drenched and frozen from the spray of the angry lake. When they passed upriver, the crowds who witnessed the drama lined the shore of the river and greeted the rescued and the rescuers with wildly enthusiastic cheers and clapping. The crowds shouted gratitude for such a magnificent, heroic rescue.

Joseph was recognized as a brave and resourceful leader. However, even though he accomplished these dramatic rescues, he was not yet a member of the U.S. Life-Saving Service, as the service did not exist until June 1871. At the time, the service was still a tiny part of the Revenue Service and did not become a separate entity until legislation passed in 1878. This service eventually merged with the Revenue and Lighthouse Services, creating the U.S. Coast Guard in 1915.

After the rescue of the *Tuscarora*'s crew, Joseph Napier was restless with the city life of Chicago and decided to move to Benton Harbor, Michigan, in 1860 to become a farmer. The quiet and profitable life appealed to him because he was now raising a young family. He moved with his wife, Anna; daughter Ada, age eight; son Clarence, age five; and daughter Sarah, age two. Shortly after moving to Benton Harbor, he decided to move across the river to St. Joseph, Michigan, following his brothers Nelson and Jack. Nelson was already popularly known as the dynamic and handsome captain of several passenger ships owned by the Goodrich line. Napier Avenue in St. Joseph was named for Nelson after he died in the sinking of the *Alpena* in 1880.

It was not until 1874 that the fledging lifesaving service decided to construct Lifeboat Station No. 6 as part of District 10 in St. Joseph, Michigan. The station was built in 1876 and put into commission on May 1, 1877. It was located at the same site as the present-day Coast Guard facility. They promptly named Joseph Napier as its first keeper on July 11, 1876.

Being a keeper of a lifeboat station, especially in the young lifesaving service, was a daunting task. Joseph Napier's men probably found him a hard taskmaster, but he had one advantage over most of his crew. He knew the mental and emotional strain involved in saving lives from the waters of Lake Michigan when they were an unbridled tempest.

On October 10, 1877, Lake Michigan demonstrated the ferocity that strikes dread into the hearts of anyone who sailed its waters. Taking on this storm was the schooner *D.G. Williams*, loaded with lumber, with her final designation in the old stomping grounds of Joseph Napier: the port of Chicago. As the schooner struggled against the relentless waves, the strain on her oak hull slowly started to allow small amounts of water to squirt into the hull. Her oakum—tar-soaked rope placed between her boards to make her watertight—began to loosen, allowing more and more water to enter her hull. At first, her crew was able to keep her hold clear of water, but as the storm grew in intensity, so did the strain on her hull. The trickle of water became a flood and overwhelmed the crew's efforts to pump it out. The captain of the *D.G. Williams* decided to be cautious and seek the shelter of St. Joseph, Michigan, only to sink on the way to Chicago.

When the *D.G. Williams* approached the harbor, the captain miscalculated and ran aground on a sandbar outside the entrance. She quickly settled on the bottom. The storm was blowing heavy gale-force winds, and waves were breaking completely over her deck, forcing the six crewmen to climb the rigging to escape the rampaging waves. Her plight was seen by those on shore, and the alarm was given to Joseph to save them.

The only problem was that Joseph hadn't had time to train a full crew for the duties of the lifesaving service. Nevertheless, he quickly got three volunteers and a boat from the propeller *Messenger*. The three volunteers were probably men Joseph was considering to help form the core of his lifesaving crew of eight.

On their first attempt to head out to the wreck, the waves capsized the boat and forced it and the men back to shore. The boat was emptied of water and valiantly pushed out to the ship for a second attempt. This time, the crew was able to make it out to the *D.G. Williams*. Because of the size of the boat, only two men could be taken off the ship at a time, and these two men were successfully landed on shore. After a short rest, Napier and his three volunteers heroically pushed off again to rescue another two men. The waves broke over the boat and completely filled it with water. One of the men bailed furiously while two men rowed, and Napier steered the boat, trying to keep it headed into the waves. After a grueling effort, they finally reached the wreck and again successfully landed two of the trapped sailors on shore.

By this time, the three men and Captain Napier were physically exhausted, but two sailors were still clinging to the rigging of the wrecked schooner. This is when Captain Napier showed the type of fortitude that gave him the honor of having WPC-1115 fast-response cutter named after him. His determination and bravery inspire the Coast Guard today.

For the fourth time, they shoved off for the wreck of the *D.G. Williams*. Exhausted and cold, with muscles aching from being soaked in cold water, they pushed onward. As they pulled close to the wreck, Lake Michigan decided to make their task more difficult. It hit the small boat with a furious surge of water and flung all four men into the air. After landing in the water, one man decided that he could swim safely to shore. Another grabbed a rope thrown to him from the last two survivors still on the *D.G. Williams* and was pulled to the wreck. Captain Napier, thrown into the air, landed in such a way that he severely injured his leg by hitting the boat as he landed in the water. Joseph and the remaining man in the water pulled the small boat close to the wreck. Then, with the help of the three men on the wreck, they were able to empty it. With the two remaining shipwreck sailors and the two volunteers, Captain Napier shoved off into the tempest to bring them all safely back to shore. Standing in the back of the boat on a badly injured leg, Captain Napier skillfully guided the boat through the growing tempest of wind and waves. Only a man with fortitude honed from previous battles with the lake could have brought these men successfully ashore. When he got back to shore, he finally allowed himself to relax, knowing full well that he put others above himself.

On May 1, 1878, the U.S. Life-Saving Service awarded Joseph Napier, keeper of Station 6 in the Life-Saving Service District 10, the first gold medal for daring gallantry in performing an audacious rescue of the crew of the schooner *D.G. Williams*. Captain Napier was also recognized for his dramatic rescue of the crew of the schooner *Merchant* when he was in Chicago. He was the first lifesaving station keeper to be recognized for his heroics and bravery.

Joseph only stayed with the U.S. Life-Saving Service for another year. Having completed three perilous rescues, he resigned as a keeper at the end of the season in 1879. At that time, there wasn't anything like welfare, life insurance, retirement benefits for workers or any type of compensation if the person providing the household's income died. The wife and children, in many cases, simply lost everything and became homeless. Also, the government paid its employees poorly, and the keepers of a lifesaving station were no exception. A station keeper, expected to risk his life saving mariners in distress, was paid less than a lighthouse

Grave marker for J.A. Napier. *Author's collection, taken at St. Joseph City Cemetery, grave Section 12, Row 1.*

keeper, who might be snug and warm in his lighthouse tower during a storm. His brother Andrew "Jack," who helped him in the *Tuscarora* rescue, died in a tragic cannon explosion during a Fourth of July celebration in 1859. His other brother, Nelson, the valiant and popular captain of the *Alpena*, went down with his ship in 1880 on Lake Michigan.

Joseph's concern for others extended to his family at home. He took those responsibilities very seriously and chose a life that was less risky than going out in storms to rescue distressed sailors. He likely was taking care of his brothers' orphans and wives along with his own family. Joseph died in 1914 at the age of eighty-eight and was buried at the St. Joseph City Cemetery. His grave is identified by a simple marker.

The next time you see a name on a ship and wonder "What's in a name?," do a little research. You might discover facts about the life of an extraordinary person—a real-life hero.

By the Narrowest of Margins

In the U.S. Life-Saving Service, the most minor details can either make you and your lifesaving crews successful or prove fatal to a shipwrecked crew and your career. This was the case of the first lifesaving crew of South Haven, Michigan. There is only a 1/16-inch difference in the diameter of a no. 7 and no. 9 rope used in a shore rescue. When the South Haven Life-Saving Station crew hastily responded to the wreck of the *City of Green Bay*, they forgot to put the thicker no. 9 rope on the beach cart, as required by regulations. Keeper Barney Cross could have been a hero rather than a villain. The mistakes that led to the deaths of six men in this rescue would not have been made public in the 1888 annual report of the U.S. Life-Saving Service. Keeper Cross might have continued his career in the service; instead, he left in shame and dishonor.

Keeper Cross was experienced in the methods, procedures and dangers of lifesaving during the perils of a shipwreck. He moved up the ranks at different stations through his bravery and skill. In 1886, he was promoted to keeper and given the command of the new station in South Haven, Michigan. His former bosses thought highly of him as brave, dependable and competent. All his former bosses recommended him to be the new station keeper there.

When he arrived at South Haven, he immediately hired, organized and trained his crew of surfmen. Glorified in the local newspapers all around the country as "Storm Warriors," they had a big reputation. Cross's obsession was readying them for the 1887 shipping season, and nothing was going to stop him from performing his duty. He pushed his crew with never-ending drills and training. The rigorous routine was performed between established beach patrols and constant lookout for mariners in trouble. When not patrolling or drilling, the men performed maintenance of the boats, gear and station grounds. The procedures were outlined in a manual provided by Sumner Kimball, the general superintendent of the U.S. Life-Saving Service. Tasks were for each day of the week. Additionally, beach patrols and the lookout tower had to be manned.

U.S. LIFESAVING SERVICE DAILY DRILLS

Day	*Drill*
Monday	Drilling for effective rescues from the beach. This included firing the Lyle gun for shooting a line out to a vessel in distress. Also, they performed maintenance on the surfboat and/or lifeboat.
Tuesday	Lifeboat and surfboat drills. If the weather was warm enough, drills were performed for capsizing the boats and righting them.
Wednesday	Practice signaling by sending and receiving messages using different combinations of flags.
Thursday	Drilling again for effecting rescues from the beach.
Friday	First-aid training and restoring the "apparently drowned" by using an early method of artificial resuscitation.
Saturday	Station and grounds maintenance and regular housekeeping and cleaning.
Sunday	Day off.

The crew's high expectations and eagerness to prove their mettle eventually turned to boredom and frustration. They began to wonder if they would ever show the local community that they were worthy of the same type of high praise that other lifesaving stations received in the newspapers. Back in 1887, there were no sports teams, movies, television or radio, so the local population usually turned to the local lifesavers to be their heroes. The person who felt the highest level of anxiety and frustration was Keeper Cross himself. Not only was he responsible for planning drills and patrols, but he also was in charge of keeping discipline and professionalism at the station.

This boredom and frustration started to cause discipline problems within the ranks of his crew.

Anyone who has had a command position will tell you that no matter how much responsibility you had when you were a step below a command position, you are not prepared for when you are the one in charge. That realization can hit a person hard. You are the person responsible for the training your crew and making them a highly proficient team. Waiting for something to happen is the hardest part for a crew. It is difficult to predict if the long hours of training and discipline will positively affect rescue performance in adverse conditions. The longer the wait, the more the stress and anxiety can wear you down. Barney Cross had waited ten long months during the 1887 season to get his first chance to lead his crew in a rescue. By then, the stress had become unbearable, and there was a high probability that he would make lethal mistakes.

The *City of Green Bay* was built in the city for which she was named in 1872. She was a stout 346-ton schooner with three masts. If she had stayed on the Great Lakes, the fifteen-year-old schooner would have been considered in the prime of her life. Unfortunately, most of her time was spent in the saltwater trade, sailing along the East Coast in the Atlantic. Salt water deteriorates a wooden ship rapidly, and saltwater animals will bore holes in the planking. In Trinidad, West Indies, a near fatal accident caused the *City of Green Bay* to be condemned by the inspectors there. Captain J.B. Hall purchased and refitted her there for trade between New Orleans and New England.

During that time, the tired and worn-out *City of Green Bay* eventually made her way back to the Great Lakes and became part of the fleet owned by A.P. Read. Salt water had damaged her wooden hull and equipment, and many people who sailed the Great Lakes thought that she needed a major overhaul. But A.P. Read was not a man to spend a lot of money based on other people's opinions. Many of the freshwater sailors would not risk their lives on such a dilapidated vessel, and Captain P.W. Costello had a tough time keeping a crew; they recognized that she could fall apart from the pounding waves on the big lakes.

The *City of Green Bay* left Escanaba, Michigan, on September 30, 1887, with 675 tons of iron ore, destined for St. Joseph, Michigan. Captain P.W. Costello and his undermanned crew of six left on a dreary afternoon and were soon in the middle of a terrific storm that would ultimately take her and several other vessels to the bottom. Lake Michigan was in a wicked mood, and by the evening of October 2, the worn-out schooner was laboring in a full gale. Contending with high waves and low visibility, she worked her way south. One of the problems with Lake Michigan, especially in those early days of sailing, was that there were not many places where a storm-tossed vessel could take shelter. The islands, where a vessel could safely shelter from the wind and waves, are all in the northern part of the lake. So, in the southern part of the lake, all the *City of Green Bay* could do is take a beating from an irritated Lake Michigan.

Recent studies have examined vessels and the strain they endure on the Great Lakes versus the oceans. Great Lakes vessels experience five to six times greater strain on their hulls than those that sail the ocean. This study surprised many people, particularly those who do not live near the lakes, who think they are just an enormous inland pond. However, the *City of Green Bay* was experiencing more stress on her tired wooden hull than what she had seen on the ocean.

She worked hard against a west wind, and her crew constantly worked her pumps to stay afloat. Today, we can start a pump just by flipping a switch or turning a key, but in the old schooners, the backs of men powered the pumps. Imagine the *City of Green Bay*'s crew standing on a heaving deck trying to move a pump lever up and down to reduce the amount of water leaking into her hull through dozens of openings. These openings were created by the twisting of the timbers against the lead caulking between the boards of her worn-out hull. The frigid waves of Lake Michigan constantly poured over the men on that pitching, wet deck, sapping their sore muscles of energy. It was pitch dark—a person could barely see the handles of the pump. The only way to know that it was working was by hearing the sound of rushing water coming out of the pump and feeling the pressure on its handle.

Anyone who has gone swimming in Lake Michigan in the summertime can attest that the lake water can be frigid; in October, the water is only a little warmer than ice. The *City of Green Bay*'s crew pumped all night, and as the darkness started changing to a very gloomy morning, they sighted the South Haven light. They were about four miles south of the light and in a position that the captain knew was too far south. It was impossible to get into the safety of South Haven's harbor. Captain Costello's only options were to try to reach St. Joseph's harbor or drop the anchors and bring her bow into the wind. Dropping the anchor might give him and his crew more time to work the pumps and hopefully reduce the rising water level inside. The boat now had more than five feet of water in her hold, and he knew that there was no chance of making it into St. Joseph's harbor. She already was loaded with twice her designed capacity, and with the extra water in her hold, she could not make it over the sandbar at the entrance of the harbor. With the raging sea and the situation on board critical, Captain Costello ordered his anchors to be dropped. In desperation, he hoisted distress signals to alert the lifesaving station in South Haven. As he waited for a sign or response from shore, the water in his hold continued to rise, and the tired *City of Green Bay* began falling apart.

Hours passed, and not seeing any help from the shore, a desperate Captain Costello made his decision to let go of the anchor chains and try to get his ship close enough to shore to give him and his crew a chance of survival. He already knew that the *City of Green Bay* would not stay afloat, but hopefully, when she sank, the ship's masts would stay high above the waves. This would give the crew a fleeting opportunity to climb into the masts for refuge from the icy water if not from the wind. Once the anchor chains were let go and with the power of the foresail, the *City of Green Bay* bounded toward shore like a tired old plow horse looking for its final resting place.

Hitting an outer sandbar with a sudden shock, the *City of Green Bay* bounced over it and swung sideways to the shore on the next sandbar. Successive shocks from the brutal waves kept her moving until her tired old hull finally came to rest on the bottom. She was 188 yards from shore (during the investigation afterward, the investigating officer from the U.S. Life-Saving Service measured the exact distance). Her crew did the next best thing to save their lives: they climbed up the masts to escape the frigid waves. Desperately, they found spots in the ship's rigging and held on, hoping that someone would rescue them.

This drama did not go unnoticed by the lifesaving station in South Haven. As the dark night sky turned from inky black to a cold, gray, overcast morning, the watch in the station's tower observed the *City of Green Bay* at anchor. He had seen her flying distress signals and riding very low in the water, with the waves washing clean over her decks. Bounding down the stairs of the watchtower, he immediately sounded the alarm and notified Keeper Cross of a ship in distress. With the alarm given, the large lifeboat at the station was readied for deployment. Keeper Cross went around the South Haven Harbor and attempted to hire a steam tug to tow the large lifeboat out to the battered schooner. Unfortunately, all the tug captains refused because of the high waves churning that cold gray morning. While Keeper Cross was looking for a steamship, the *City of Green Bay* slipped her anchor and became firmly beached in the surf offshore. Cross, frustrated by the turn of events and the lack of help from the tug captains, decided that he would try to rescue the men on the ship by means of breeches' buoy. Since the beach

Breeches' buoy. *Michigan City Public Library Collection.*

across from the wreck was too small to launch a surfboat, a breeches' buoy was his only option.

A surfboat is a smaller version of a lifeboat and is easy to transport across land, but it needs ample beach to successfully launch into the waves. A breeches' buoy is another method, but the lifesavers had to fire a rope out to the ship. The shipwrecked crew would rig ropes between the shipwreck and shore. The rope is the guide on which the buoy is run to haul the men from the ship to shore. The crew literally uses a buoy with a pair of breeches sewn to it to save a shipwreck crew.

At this point, the situation was unraveling for Keeper Cross and his valiant crew. Cross decided that he would only take the beach cart with the essentials for the rescue where the *City of Green Bay* had wrecked. The decision to leave the lighter, smaller surfboat at the station proved fatal to the *City of Green Bay*'s crew. For some reason, the beach cart also only had the smaller, lighter no. 7 rope. The wider no. 9 rope required by the service was left behind in all the excitement. Beach carts are pulled either by the crew or by horses. Stations were not allowed to keep horses, per U.S. Life-Saving Service regulations, so most keepers contracted with a local stable to always have a team of horses available. After hitching the team of horses to the cart, it was a three-mile haul to the beach opposite the wreck. Because of the size of the cart, there was only one bridge that could get the cart over the Black River in South Haven. There was a more direct route, but as it was not suitable for the horses and cart, Keeper Cross made his next mistake: he sent some of his crew members on that route to the beach opposite the wreck and signaled the people on the *City of Green Bay* that help was on its way. It took the beach cart and horses almost two hours to cover the three miles to the beach. A tremendous physical effort went into moving that heavy cart because the trail was littered with debris, branches and fallen trees from the gale. The extra crew sent on the shorter route could have helped in clearing much of this debris out of the cart's way. Once they got to the beach, Keeper Cross realized his first mistake when he saw more than twenty-eight feet of exposed beach. If they had the surfboat, they could have easily launched it and rescued the crew.

A Lyle gun was now Cross's only way to make a safe rescue. This small cannon shoots a rope line out to the wreck and brings the survivors to shore by rigging the breeches' buoy. A Lyle gun was a small bronze cannon developed by First Lieutenant David A. Lyle of the Army Board of Ordnance. When the U.S. Life-Saving Service was first established, Sumner Kimball, the general superintendent of the service, tasked David Lyle, an ordnance officer, with figuring out an efficient method of shooting a line to a shipwreck offshore. David Lyle conducted a lot of research and testing and found that the best way was to use a small bronze cannon with a two-and-a-half-inch bore and an iron projectile. It could fire the projectile with a no. 9 rope attached more than 700 yards (2,100 feet). A typical powder charge for this cannon was one and a half ounces, but it could take up to eight ounces of powder. With one and a half ounces, the cannon would recoil back six feet. With eight ounces, the cannon's recoil would be extremely dangerous, and the cannon and its mount could be damaged. David Lyle thought that eight ounces of powder was too large of a charge and hazardous to those operating the cannon. One of the advantages of the Lyle gun was its small size. Two men could easily carry it. However, the size was also a disadvantage during the cannon operation because of the dangerous recoil from its light weight, especially when the powder charge exceeded more than the recommended amount. The size and strength of the rope attached to the projectile was also critical because if the amount of powder were too great, the muzzle flash and shock could

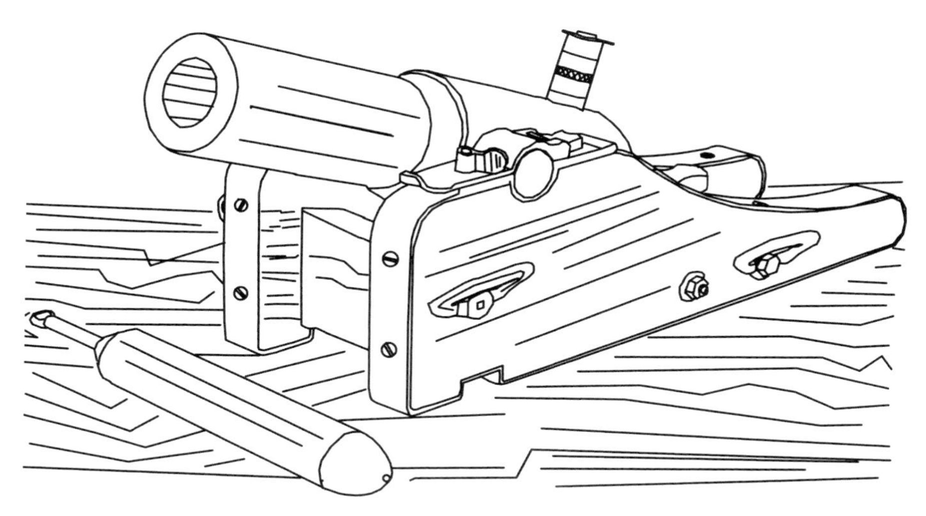

Lyle gun line art. *Created by the author.*

break the rope. Also, the first shot is crucial because the rope's weight can change dramatically when wet. This reduces the range of the Lyle gun and increases the amount of the powder needed on successive firings of the projectile.

With the noise of the surf raging in their ears and the sight of the wreckage from the *City of Green Bay* coming ashore, the lifesavers hastily lowered the gun down an embankment, about 30 feet, and prepared to fire the rope out to the wreck. One and a half ounces of powder would enable it to shoot a line out 700 yards, and the target was only 188 yards away from shore. Cross's men saw that the waves were sweeping over the shipwreck, and her crew had escaped by climbing into the rigging. Two masts of the *City of Green Bay* were still standing, and the crew's weight on the rigging was the only thing that held the masts in place. However, it also caused them to sway perceptibly. They charged the gun with more than one and a half ounces of powder, inserted the projectile down the muzzle and fastened the no. 7 line to the projectile. Keeper Cross aimed carefully and fired. He saw the projectile sail right between the two masts, and at the same time, the no. 7 rope broke from the projectile about fifty feet into its trajectory. He ordered the now wet line to be brought back in and recoiled for a second try. Reloading the gun, he aimed a second time and fired. This brought the same results as the first try. The beach cart carried only three projectiles, and he had just lost two due to the weaker no. 7 line and the heavier charge of gunpowder. He ordered two of his surfmen to go back to the station with the team of horses and get the surfboat, more projectiles and the stronger no. 9 rope.

After the two surfmen left, Keeper Cross had the same no. 7 rope, which had already broken twice, recoiled and connected to his remaining projectile. By this time, the no. 7 line was thoroughly soaked from being fired into the waves twice, which made it even heavier and more prone to breakage. He fired it a third time with the same results as the first two attempts. Now he was out of options. He didn't have the right rope or the surfboat, and the ship's crew was growing weaker from exposure to the cold, wet wind. In desperation, Keeper Cross hurried after the two men sent to retrieve the surfboat. After about forty-five minutes, one of the surfmen returned with extra projectiles and no. 9 rope. The number one surfman of the crew loaded the Lyle gun with four ounces of powder and attached the new rope to the projectile. The shot was accurate and landed over the cross-tree of the mast. Just when it seemed that things were about to go right, bad fortune reappeared. Two of the ship's crew had climbed down from the main mast and gone to the forward mast to retrieve the no. 9 rope. While they were climbing the mast, it fell over into the surf, drowning both unfortunate sailors. Not only were two sailors lost, but now the no. 9 line was too.

The remaining five sailors, still in the rigging on the aft mast, began to fear that it was going to fall over too. So, they climbed down onto the wave-ravaged deck to find whatever shelter they could. By then, the surfboat had arrived and was launched. The lifesaving crew began pulling hard toward the wreck, but their progress was impeded by the wreckage that came off the ship. With heroic and desperate strength, they pulled toward their goal. Again, as they neared the boat, they witnessed men being washed off the wreck and drowned. Only one of the crew of the *City of Green Bay* was saved. He desperately hung on to wreckage, and when his strength finally reached its limit, the waves washed him close to the surfboat, where he was pulled aboard. As the despondent lifesavers turned the surfboat back toward the beach, they spotted Captain Costello's body floating in the waves. They brought his body into the boat and tried to resuscitate him, but he was already too far gone.

Summer Kimball of the U.S. Life-Saving Service took great pride in the lives the service had saved over the course of a year. He developed the service into a professional and exacting organization. There was minimal leeway when a life was lost during a rescue. So, when the report was filed concerning the wreck of the *City of Green Bay* and the six who died, the service sent an investigator. The investigator meticulously questioned the sole survivor and those present during the rescue, including the lifesaving crew and spectators. He examined the equipment's condition and found that the carriage of the Lyle gun was split in several places, indicating that too much powder was used firing it. He test-fired the cannon with the recommended amount of powder and a no. 7 rope. The projectile went more than four hundred yards with the same no. 7 rope used in the rescue. This showed the investigator that too much powder and too small of a rope were used. His final report blamed poor Keeper Cross and his inability to make sound judgments under stress. This was a harsh conclusion, especially considering the recommendations and accolades Keeper Cross had received. The report stated, "A good soldier does not always make a good captain." No matter how well he performed at other stations, Keeper Cross was responsible for the mistakes made during the unsuccessful crew rescue of the *City of Green Bay*. Barney A. Cross resigned from the U.S. Life-Saving Service on December 21, 1887, a little more than a year after taking charge in South Haven. If he had used a no. 9 rope, he may have been considered a hero. However, the poor decisions made and the outcome of the rescue placed the blame squarely on Keeper Barney Cross.

Same Bloody Storm

Captain John Curran was looking over the *Havana* and thinking how lucky he was to be her captain, instead of her dock partner and fleet mate the *City of Green Bay*. Even though the *City of Green Bay* was bigger and a year younger than *Havana*, the *Havana* was in better overall shape. The *City of Green Bay* had spent too much time in the saltwater trade. Her equipment, hull and everything about her showed the detrimental effects of salt water on a wooden vessel.

As both boats sat at the dock on Thursday, September 29, 1887, loading iron ore in Escanaba, Michigan, both were heading to St. Joseph, Michigan, with their cargo. There were similarities between these two vessels: They were both wooden schooners owned by the same person, Alonzo P. Read. Both ships were loaded with iron ore and planned to leave for St. Joseph, Michigan, on September 30 to unload their cargo. But a primary difference between the *Havana* and the *City of Green Bay* was that the *Havana* had two masts, while the *City of Green Bay* had three.

Captain Curran felt that he had the better vessel and crew. He appreciated how his ship stood on the water under full sail like a graceful lady in the prime of her life. She was a pleasure to sail, and his crew, which had been with him all season, could sail her smartly. Captain Curran was in his first command for about a year, and his crew was about the best you could get in 1887. Furthermore, having the same crew members all season was very unusual for any vessel. The crew was as follows:

Schooner *Havana*'s Crew

Crew	*Rank*
Sam McClement	First Mate
John Morse	Steward
Joseph Clint	Sailor (cousin to Captain Curran)
Charles Hagen	Sailor
George Hughes	Sailor
Robert McCormick	Sailor

All these men came from Chicago, except for John Morse, who hailed from Benton Harbor. John Curran had immigrated to America from Ireland in 1872 with his father, James Curran, and his three brothers, James, Thomas and Robert. His father, like all his brothers, sailed the Great Lakes. John worked on several different vessels and finally became the first mate on the schooner *Ada Medora*. He served as mate until he was given command of the *Havana* in 1886.

Even though it was late in the season on that cloudy Thursday in September, Captain Curran was looking forward to the trip down Lake Michigan. The wind was out of the north-northeast, and he expected an easy sail to St. Joseph. Little did he know that this was going to be his last trip.

The *Havana* was built in Oswego, New York, for Thomas S. Mott's fleet of ships and launched in August 1871. Hundreds of spectators crowded the shipyard that hot and sweaty afternoon to see her touch fresh water for the first time. The spectators bustled around as the workers made final preparations for that blistering afternoon launch. Finally, everything was ready, and the workers, drenched in sweat, got the signal to knock out the last blocks that were holding her back from sliding into the water. The majestic lady they had slaved over for the better part of a year gracefully slid into the water. There were enthusiastic cheers from the crowd as she tasted fresh water for the first time. The small army of workers that built her bristled with pride as they saw their creation float for the first time. The launch was nearly perfect in every detail, and to the many superstitious sailors watching the event, that meant she was going to have a long and profitable life. With the *Havana* in the water, it was time to celebrate and toast the launching of a fine ship.

She sailed the freshwater seas for sixteen years, with only one major repair in 1883. Then, in the fall of 1887, A.P. Read needed another ship to expand his small fleet of three ships. His three vessels were overbooked, and he needed help meeting the increasing demand for high-grade iron ore. Even though not considered a large schooner, the *Havana* could safely haul 306 tons of ore. He put her right to work, fully expecting a tidy little profit from the remaining months of sailing on the lakes in 1887. Little did A.P. Read realize, at the time, that fate had other ideas for his little fleet. By the end of October, three of his four ships would be on the bottom of Lake Michigan, and his fourth ship would be severely damaged.

Leaving Escanaba, Michigan, with 551 tons of ore, several hours ahead of the *City of Green Bay*, *Havana* pointed her bow south toward the port of St. Joseph, Michigan. The northwest wind was ideal. She was smartly sailed and made good time. By the afternoon of October 1, Lake Michigan had started to brew an enormous gale of a storm. As the winds increased, she began to take a pounding but handled the inclement weather with very little fuss. However, by the evening of October 2, she was in a full-blown gale, and getting over the sandbar safely to St. Joseph weighed heavily on Captain Curran's mind. He knew that the only safe way to enter the harbor, especially in a storm with an overloaded vessel, was with help from one of the local tugs. So, he picked a place about a mile out in the big lake, dropped his anchors and ran a signal up his mast for a tug to come out to help him get the *Havana* safely into the harbor.

Until now, the *Havana* and her crew were not having difficulty keeping the hold's water level in check. She was equipped with two man-powered pumps, one near the ship's bow and the main pump near amidships. As time passed, the fury of the storm increased, and so did the waves. They began to wash over her amidships and allowed more water into her hold. This prevented the crew from using the larger-capacity main pump. The smaller pump near the bow had some shelter from the waves but had to be worked faster and harder to keep the water below from rising. The violent waves were not her only problem. The seams in her hull also started to open, increasing the amount of water entering. The men who were not operating the pump began to desperately use buckets to aid the men at the pump.

The old-time wooden sailing vessels had to have their seams (the space between the wooden boards) caulked, meaning that oakum, cotton or rope fibers were driven into the gaps between the boards. This protection helps make the hull and deck watertight.

The caulking process was complicated and dependent on the skill of the person performing it. Over time, caulking rots and needs to be replaced. Replacing the old caulking will keep the ship watertight, but this process was expensive. Most ship owners could not bear that expense, especially for an older vessel. A.P. Reed was not a man who would spend a lot of money on repairs that were not obviously necessary. The ship might not leak when it was sitting in calm waters, but in a storm, with her hull support and planking bending and twisting, the caulking would often fail, allowing streams of water to enter the hull. The old-time sailors called this type of leaking "spitting her caulk." The longer the old wooden ships were subjected to storm stress, the more they leaked.

After she dropped her anchors, the crew struggled to keep the ship pumped out, hoping that they would get a signal from those on shore that a tug was coming. Around 7:00 a.m. that Sunday, her crew was exhausted from constantly working the pump and using buckets. Captain Curran was out of options and decided that the only chance of survival for him and his crew was to try to maneuver his boat onto the beach. This would give his crew the best survival odds and allow the salvage of the cargo after the storm subsided. He ordered her anchors slipped and the after-sail set, and the *Havana* bounded north, away from the harbor and parallel to the shore.

This drama caught the attention of those on shore. When the lookout in the tower at the St. Joseph Life-Saving Station noticed the *Havana*, she was still at anchor outside the harbor with distress flags flying. Unfortunately, Captain William Stevens of the St. Joseph lifesaving unit was already working on the small schooner *Harvey Ransom* of South Haven, Michigan. At the outset of the storm, the *Ransom* decided that it was best to run for shelter in St. Joseph's harbor. The *Ransom* was a very small ship—59 feet long and 16.33 feet wide. She was much too small to endure the storm raging on Lake Michigan. As the *Ransom* approached the entrance, her rudder post broke, causing her to miss the opening. Colliding with the south pier, she ended up on Silver Beach, just south of the harbor entrance. The lifesaving crew immediately launched a surfboat and rowed across the St. Joseph River to the south breakwater. They chased after the *Ransom*, and by the time they reached her, she was well up on the beach. The lifesaving crew boarded her and helped her two-man crew to furl the sails and batten down everything before making the *Ransom*'s crew leave her until the storm blew out. The *Ransom* was refloated several days later with very little additional damage and could complete her trip to Chicago after her rudder post was repaired.

While the lifesaving crew was finishing with the *Ransom*, Captain Stevens went around the harbor to enlist the help of a tug. All declined his request, stating that it was too rough on the lake to risk their vessels—all except for the fishing tug *Hannah Sullivan*. The *Hannah Sullivan* was one of the strongest fishing tugs in the harbor, and its captain, Charles Mollhagen, volunteered to risk his ship to help. The problem with the *Hannah Sullivan* was that some of her machinery had been taken apart for maintenance and needed to be reassembled to fire up the boilers. Captain Mollhagen instructed engineer August Kuehn to put the machinery back together and ready her boilers for a heavy sea.

Meanwhile, Captain Stevens made his way back to the lifesaving station, only to find out from the lookout that the *Havana* had just slipped her anchor and was heading north. Quickly, he ordered his men, who had just returned from the *Ransom*, to assemble the beach cart, and he sent one crewman to find a team of horses to pull it. Not waiting for the horses, a group of citizens volunteered to pull the beach cart in the direction of the *Havana*. Using extreme effort, they managed to pull the equipment nearly four miles, at which point the team of horses caught up. The horses pulled the beach cart another mile north, and the volunteers spotted the masts of the *Havana* sticking out of the water.

After Captain Curran slipped her anchors, the *Havana* sluggishly moved by only using her forward sail and headed north for about two hours. Five miles north of St. Joseph and about a mile short of the beach, the water level in her hold finally overcame her exhausted men, and she settled in forty-five feet of water. With the heavy load of iron ore, she dropped to the bottom like a rock, taking her crew with her. They managed to come to the surface by climbing the mast to get out of the waves. Captain Curran and two of his men climbed up the mainmast, and the remaining four crewmen clambered up the foremast. Shortly after the *Havana* hit bottom, the mainmast swayed heavily and then snapped, throwing Captain Curran and the two crewmen into the icy water. They became tangled in the floating debris and rigging and slowly drowned as the waves and debris dragged them under the turbulent water. The men frozen on the foremast watched in utter horror, only able to witness the loss of their shipmates. They wondered if they would share the same dreadful fate. In their foggy, exhausted minds, the one-mile swim to shore was a tempting siren, urging them to plunge into the freezing water, even though they would risk a slow, painful suicide. They hung to a tenuous perch on the mast, swaying with each wave, waiting for it to snap and send them to join the fate of their shipmates.

After the drama of the mainmast breaking, Captain Stevens and his crew arrived on the beach and could see the foremast still standing. Unfortunately, at a mile away, they could not see the survivors on the mast. They debated whether it would be wise to launch the surfboat and risk the lives of the lifesavers if there weren't any survivors from *Havana*. About this time, the *Hannah Sullivan* could be seen in the distance, making her way to the *Havana*.

When engineer August Kuehn was told to get the machinery of the *Sullivan* back together, word had spread around town that the fishing tug was going to attempt to save crewmen off a schooner in distress. Several citizens of St. Joseph came to the docks to try to help. Besides Captain Mollhagen and engineer August Kuehn, seven other men offered their help:

Tug *Hannah Sullivan* Additional Crew
August Habel
Robert Mollhagen
Alexander Cran
Louis Mollhagen
George W. Schneider
John Carrow
J.H. Langley

There was a discussion before leaving the safety of the harbor on whether the *Hannah Sullivan* could make it over the sandbar at the harbor's entrance. This was a significant concern to Captain Mollhagen because if she got stuck on the bar, there would be little chance to escape and all would likely perish. This would leave his family, as well as his partner engineer Kuehn, destitute. In 1887, the government did not provide any kind of welfare or food stamps for families. When they left, the crew was unsure if anyone on the *Havana* was alive, but they were determined to find out and rescue anyone who was. As stated in the Lifesaving Report of 1889, "The men were large-hearted and resolute and determined to make the attempt."

After successfully breasting the waves over the sandbar at the harbor entrance, the *Hannah Sullivan* traveled eight miles of chaotic wave-tossed lake to get downwind to the *Havana*. The spectators on shore who watched the fishing tug go out into the storm doubted that anyone on the ship remained alive. There was a chorus of prayers for both the men on the tug and anybody left on the *Havana*.

The men on the *Hannah Sullivan* finally spotted the *Havana*'s lone mast sticking out of the water with a fragment of the topsail flapping. They could see four men still clinging to life on the mast, along with debris of rope and broken wood from the *Havana*. Any piece of rope or wood was likely to get tangled in the *Hannah Sullivan*'s single propeller and end all hopes of rescue. The tug steamed to the wreck's windward side, and Captain Mollhagen moved the *Hannah Sullivan* as close as he dared. With the ample amount of manpower on board, he decided to put a small boat in the water to make the final approach. A rope was fastened to the boat to make it easier for those on board to be pulled back after they rescued the crew of the *Havana.* As they pulled near the mast, three men were able to scale down the mast and get into the boat without much assistance. The fourth man, First Mate Sam McClement, was so far gone with exhaustion and hyperthermia that he did not make it then. After bringing the first three men aboard, Captain Mollhagen made another approach to the mast, and a line was heaved to the exhausted first mate. Mate McClement took the line, tied it around his waist and then fell into the water. The men hauled him on board, more dead than alive. Bundled quickly in a blanket, he was taken below the deck to warm up near the boiler. Thankfully, after a strong drink, he revived.

Once the four survivors were on board the *Hannah Sullivan*, a quick search ensued for the three men lost when the mainmast snapped off. When Captain Mollhagen was satisfied that nothing more could be done, he headed the *Hannah Sullivan* back into the storm-tossed waves and toward home in the St. Joseph Harbor. Still fearing the sandbar at the entrance, he cautiously approached, and with the skill and teamwork of him and his engineer, he got the *Hannah Sullivan* back into the harbor.

After seeing the *Hannah Sullivan* leave the *Havana*, Captain Stevens assumed that survivors, if any, were safely aboard and that the services of the lifesaving team were no longer needed. He ordered his crew to pack up all the gear and get back to the station. He knew that his services might still be required if the *Hannah Sullivan* failed to cross the dreaded sandbar. Once they got back to the station, they found that Captain Mollhagen was already back in the harbor and that the four survivors were being treated for exposure by a local doctor. After a brief time with the doctor, First Mate Sam McClement, Charles Hagen, George Hughes and Robert McCormick boarded a train for their hometown of Chicago.

When they arrived in Chicago, Sam McClement was interviewed by reporters, and he blasted the lifesaving station in St. Joseph for not having acted promptly. He blamed Captain Stevens and his crew's inactions for the deaths of Captain Curran, John Morris and the captain's cousin, Joseph Clint. McClement stated that if the lifesaving crew in St. Joseph had quickly responded when the *Havana* put up distress flags, the crew would have survived. He did not know that the lifesaving crew was busy trying to rescue the schooner *Harvey Ransom*. McClement also did not realize that Captain Stevens had asked for the *Hannah Sullivan* to help take his lifeboat out to the *Havana* but that the *Havana* had slipped her anchor, and the lifesaving crew was chasing her down the beach. The only reason the *Hannah Sullivan* was able to retrieve the survivors of the *Havana* was that Captain Stevens had asked Captain Mollhagen if he could get the *Hannah Sullivan* ready to go out in the gale. Only when Captain Mollhagen found out that the lifesavers were already heading down the beach after the *Havana* did he go out in the storm and try to rescue the survivors. Sam McClement did not stay long enough in St. Joseph to find out all the events involved in his rescue. All he knew was that he had watched his captain drown and needed someone to blame. Unfortunately, he blamed the lifesavers.

In 1889, the U.S. Life-Saving Service decided to reward the crew of the *Hannah Sullivan* for their gallantry and bravery in saving other men's lives. Gold medals were awarded to Captain Mollhagen and engineer Kuehn for their noble, selfless actions. Silver medals were awarded to rest of the crew: John Carrow, J.H. Langley, August Habel, Robert Mollhagen, Alexander Cran, Louis Mollhagen and George Schneider.

The perilous November weather has always made both ship captains and owners nervous. November storms on the Great Lakes have become legendary. Storms like the 1905 November storm, called the "*Mataafa* storm," occurred late in November and claimed the steamer *Mataafa* at the entrance to the Duluth Harbor in Lake Superior. That storm also claimed nineteen other ships. To avoid future tragedies, the Split Rock Lighthouse was built north of Duluth. Other legendary storms include the 1913 Storm, sometimes called the "Big Blow," where each of the five Great Lakes except for Lake Ontario lost a ship. Twelve ships sunk, eight of them on Lake Huron alone. They accounted for the deaths of more than 244 sailors, not including the fatalities on shore. Also, there was the Armistice Day Blizzard of 1940 and the more recent Fitzgerald Storm of 1975, during which the *Edmund Fitzgerald* went down on Lake Superior, taking all twenty-nine of her crewmen with her.

These legendary storms demonstrate the ferocity of the Great Lakes. So, A.P. Read must have felt extreme anguish when the telegraphs arrived in his office reporting the loss of the *Havana* and the *City of Green Bay*. Not only were these incidents huge financial losses, but also almost two full crews and two experienced captains perished. As A.P. Read paced his office, hoping that some good news would come, he must have kept muttering that half of his fleet had been wiped out in the same bloody storm.

One More

Richard Stines sat shivering on a cold, gray, stormy morning in the St. Joseph Life-Saving Station tower. He hated being exposed to the cold wind blowing off the gloomy waters of Lake Michigan. The morning of November 11, 1887, with dark, low-hanging clouds, was particularly miserable. He would rather be on the beach patrol. At least he could keep warmer by constantly walking along the beach. A ghostly, dim outline of a schooner suddenly appeared in the churning lake. As it came closer, he saw three masts. Familiar with the schooners that came into St. Joseph, Stines knew it was not the *Havana* or the *City of Green Bay*. Those schooners went to the bottom of Lake Michigan earlier that month. As she came closer, he was pretty sure that it was the *Myosotis*. Even in the gray morning, he could tell that she was sitting low in the water and was probably overloaded. He wondered what kind of a fool captain would load his ship until it sat that low in the water, especially in the stormy month of November. It barely had enough freeboard to keep her afloat. This was a time before insurance companies enforced load lines and safe loading limits on the ships.

With the wind and spray coming off the lake, it was hard to keep the *Myosotis* in view. As she drew closer to the harbor's entrance, Stines saw more details and concluded that she was loaded beyond a safe limit. His concentration intensified as the *Myosotis* approached the sandbar near the entrance of the harbor. He did not expect her to make it over the sandbar, which was already responsible for the loss of the *Havana*, one of the *Myosotis*'s fleet mates. The *Havana*, also overloaded, could not make it over that sandbar without help from the local tugs. She sat outside the harbor, waiting in vain for a tug to rescue her. But the stormy waters of Lake Michigan loosened her seams, and she started to leak badly. The only option was to try to beach her, and she sank, taking three crew members with her. Also, that sandbar was probably a contributing factor when another fleet mate, the *City of Green Bay*, dropped its anchor much farther north of St. Joseph, only to get pounded to bits and pieces in the surf. The *City of Green Bay* lost six of her seven crewmen to drowning and exposure.

Suddenly, as if struck by an invisible hammer, Stines's attention snapped back to the *Myosotis* as she was struck by a huge wave and dropped onto the sandbar. Several more waves slammed her again, and each collision was fatal. As he helplessly watched, another series of waves lifted the ship and beat her again on the sand bottom, as if some demon was playing with her. With damage to her rudder, centerboard and steering, the *Myosotis* spun completely out of control. As she moved backward, her stern pointed toward the entrance, and at the mercy of the wind and waves, she headed for the south breakwater. As if the pounding on the sandbar was not damaging the *Myosotis* enough, another resounding crash exploded through the gray cold morning, as she hit the end of the southern breakwater. Fatally damaged, her shattered hull was pushed south toward Silver Beach. Stines could clearly see the men on her deck desperately running to prepare for the worst. He frantically raced down the tower's stairs and rang the alarm to summon the lifesaving crew to the *Myosotis*.

The *Myosotis sylvatica* plant is one of the most delicate flowers in the state of Wisconsin. The flower itself can be five blue or white petals with a yellow center. The name *Myosotis* refers to the Latin term *mys*, meaning mouse, and *ous*, meaning ear, and refers to the flower's delicate petals. The ship must have been a thing of beauty to her builders and owners.

Hull no. 90764 started its life in 1874 at the Milwaukee Shipyard in Wisconsin. Allen and McClelland, her original owners, wanted a ship that could haul cargo for their fleet. They specified a three-mast schooner with a length of 134 feet and a 24-foot beam (width). She could draw 14 feet of depth with a full load of 316 tons of cargo. Made with the finest oak and lumber from a Wisconsin forest, the hull of the *Myosotis* took shape with the shipwrights' sweat and toil. When she was launched, the local newspapers extolled her as one of the most elegant schooners on the Great Lakes, much like the native Wisconsin flower that bears her name. She was enrolled on May 26, 1875, and put to work by Blyer Brothers of Milwaukee. The *Myosotis*

The *Myosotis*. *Alpena County George N. Fletcher Public Library, Great Lakes Maritime Collection.*

was sold later, becoming an addition to Alonzo P. Read's small fleet. He immediately put her to work hauling iron ore from Escanaba to St. Joseph for the growing steel industry.

The *Myosotis* started down Lake Michigan from Escanaba to her final designation: Spring Lake Iron Company in St. Joseph, Michigan. She was loaded with six hundred tons of high-grade iron ore. Like with the *City of Green Bay* and the *Havana*, A.P. Read had loaded the *Myosotis* to exceed her maximum cargo limit. She met unfavorable winds coming down the lake, which forced Captain John Maloney to put her into Chicago on Tuesday, November 8, 1875. Two days later, the weather turned into a beautiful November day, and Captain Maloney sailed the *Myosotis* out of Chicago around 5:00 p.m. He expected the fifty-nine-mile evening sail across the lake to be an enjoyable trip with favorable winds at his back. But any sailor will tell you that the Great Lakes in November can be unpredictable. During the night, the wind shifted to the northwest and turned the almost smooth surface of the lake into choppy, wind-driven waves. Since the ship was about halfway across the lake, it was out of the question to turn around and go back toward Chicago. Captain Maloney had no choice but to keep going toward St. Joseph.

William Stevens, captain/keeper of the lifesaving station in St. Joseph, Michigan, heard the alarm from Richard Stines and hurried his crew to launch the lifeboat from the station. Hours of training paid off as his crew prepared the boat to rescue the *Myosotis* crew. With precision, the lifesavers manned the lifeboat, and it slid down the ramp into the channel. Pulling hard, the men could already feel the rise and fall of the rough waves. Captain Stevens expertly guided the boat perpendicular to the incoming waves. At the entrance to the channel, he carefully timed the boat's turn to the south, so the amount of time it spent in the trough of the waves would be minimal. Once around the end of St. Joseph's south breakwater, he could see that the *Myosotis* had already ground itself into the bottom, just off the shore. The small boat that usually hung on her stern was missing. In the pitch-black night, he spied the *Myosotis* crew rowing the small boat toward shore and wondered if they would make it through the dangerous breakers. If any of those men got washed into the lake, their survival time in the icy waters of Lake Michigan in November would be measured in minutes. Yelling over the roar of the wind, he urged his crew to go faster in case the small boat would be swamped and men lost. His men were trained for the conditions present on the lake, but typical sailors on merchant vessels like the *Myosotis* were not trained to handle a small boat in the rough waters. As

St. Joseph Life-Saving Station and crew. *Maud Preston Library Collection, St. Joseph, Michigan.*

they closed the distance, Captain Stevens could see the *Myosotis*'s small boat enter the breakers, and he held his breath as the boat's stern dramatically rose on the first wave. Captain Stevens's heart raced with each successive wave, wondering if he was going to witness a disaster. He finally saw and heard the small boat crunch on the sand on the beach. He and his crew skillfully brought the lifeboat next to the *Myosotis*. The lifesavers jumped out and hurriedly helped the *Myosotis* crew pull their boat out of the reach of the breakers on the beach and made sure that their boat was also safe.

Captain Stevens and his men quickly checked Captain Maloney and his crew for injuries or any need for assistance. Other than being cold, wet and tired from their battle with the chaotic waves of Lake Michigan, the men of the *Myosotis* did not need any help. So, they were led back to the lifesaving station, where Captain Stevens's wife, Ella, prepared a hot meal. With dry clothing and a warm bed for the evening, the *Myosotis* crew left the station the following day around 7:00 a.m.

The next few days, the lifesaving crew helped row the *Myosotis* crew back to the ship to get their personal possessions. The ship was in shambles, and Captain Maloney had to report to A.P. Read that he had lost another one of his ships that month. The waves had smashed in the cabins, and the

Myosotis in ice. *Alpena County George N. Fletcher Public Library, Great Lakes Maritime Collection.*

hull was crushed. The crew found only a small portion of their personal clothing and gear. An agent representing the underwriters for the *Myosotis* came from Chicago and hired a local contractor to salvage the vessel. By Sunday, November 13, the contractor had pulled off the schooner's rigging and anything valuable and stored it in one of the Graham & Morton's warehouses in Benton Harbor.

November 1887 was a disastrous month for Alonzo Read. Three of his four-ship fleet—the *City of Green Bay*, *Havana* and now the *Myosotis*—had been lost. His remaining ship, the *D.S. Austin*, sustained $12,000 worth of damage, financially wiping out A.P. Read's company. He eventually recovered and went back into business. He moved from Kenosha, Wisconsin, to Chicago, Illinois, where he died at the age of seventy-four in 1915. It makes a person wonder if, over his lifetime, he ever felt regret in overloading his vessels—actions that led to his bankruptcy and the deaths of the men who had sailed for him.

Only Grit

The dramatic 2016 movie *The Finest Hours* portrays a Coast Guard crew on the U.S. East Coast that went out in a killer storm in 1952 to perform a daring and perilous rescue of a crew from a Liberty ship called the *Pendleton*. The *Pendleton* had split in two, with her bow section sinking, and thirty-eight crewmen on the ship's stern. They were in desperate shape and needed to be rescued. Watching that movie, you can't help wondering how those Coast Guard men possessed that kind of grit. They risked everything to save a shipwrecked crew. What was even more remarkable was that they undertook the mission in a full-gale winter storm, going out in a thirty-six-foot boat with an open cockpit. It exposed the crew not only to blizzard conditions but also to freezing twenty-five-to-forty-foot waves.

Grit is defined as "courage and resolve; strength of character in a person." This definition aptly describes the men in the *Pendleton* rescue and other lifesaving and Coast Guard crews that performed amazing rescues. But what about a crew that went out in a killer storm and did not save any lives but still performed an extraordinary feat of survival? Do they have grit? Only you, the reader, can decide if they possessed the same resolve and strength of character as if they were successful.

Twelve years before the killer storm on the east coast, another raged on November 11, 1940. The storm was not in the Atlantic Ocean but rather on Lake Michigan. Another courageous Coast Guard crew set out into this storm from South Haven, Michigan, to rescue not the crew of a Liberty ship but rather two separate crews on commercial fishing tugs. The crews of

Thirty-six-foot lifesaving boat. *Michigan Maritime Museum-South Haven Michigan.*

those tugs were friends and neighbors of the Coast Guard men. The fishing tugs had set out early on that fateful day to take advantage of the beginning of the fishing season and the unusually warm fall Indian summer weather. This killer storm became known as the Armistice Day Storm of 1940. It claimed the two fishing tugs and the lives of 145 people, 66 of whom were the crews of two steel steamers, the 308-foot SS *Anna C. Minch* and the 420-foot SS *William B. Davock*. That number included two men off the 250-foot SS *Novadoc*. The remaining people were unfortunate duck hunters and civilians in Wisconsin, Illinois, and the crews on the fishing tugs. This Coast Guard crew was not as widely known in the service as those men on the east coast because the fishing tugs and their crews perished on Lake Michigan. Even though no one was saved, it still took a huge amount of courage to face the towering 35-foot waves of Lake Michigan for an incredible twenty-five hours. The fishing tugs from South Haven, Michigan, were the *Richard H.* and the *Indian*.

The *Richard H.* had a crew of three and was owned by Captain John McKay. He was not on the *Richard H.* when she sank, but his twenty-eight-year-old son, John Jr., was the captain of the tug when she went out that fateful morning. The two other crew members were thirty-five-year-old John Taylor and thirty-three-year-old Stanley White. This was the first trip Stanley White took on the *Richard H.* and sadly, his last.

The *Richard H.* was a brute nineteen-ton steam-powered tug, 43.8 feet long with a 12-foot beam. Built in Marinette, Wisconsin, in 1923, she was considered a tough and sturdy workboat. However, her steam power plant was not as dependable as the more modern diesel power plants. The early steam plants were difficult to keep going in rough conditions because their coal or wood fuel had to be shoveled by hand into the boiler. The *Richard H.* left South Haven around 8:00 a.m. to set nets for lake trout, salmon and whitefish. The fishing season had started the day before the storm hit.

The other tug from South Haven was the *Indian*. The *Indian* was an older tug built in Manitowoc, Wisconsin, in 1914. She was 40.1 feet long and 12 feet wide. Four years before, in 1936, the *Indian* had a more reliable Kahlenberg diesel engine installed, which replaced her outdated steam engine. Diesel engines did not require the constant attention and care that steam propulsion needed.

The *Indian* had a four-member crew when she coasted out of the South Haven piers that fateful morning: fifty-five-year-old Captain James Madsen, thirty-five-year-old Harold Richter and thirty-year-olds Bill Bird and Art Reeves. The *Indian* left South Haven at about 7:35 a.m., a little earlier than the *Richard H.* Like the *Richard H.*, they planned an all-day fishing excursion.

No one expected the terror that was moving toward them from the west. Earlier that week, the storm had collapsed the Tacoma Narrows Bridge in the state of Washington and froze hundreds of cattle and livestock in the fields of other western states. Weather reporting in 1940 was a lot different than today. The National Weather Service offices were only open twelve to fourteen hours a day, reporting the weather daily at 8:00 a.m. and 8:00 p.m. The observations were based on ground-level readings and telegraphed to the main office in Chicago, Illinois. There were no weather maps, and weather radar did not exist. Additionally, there was no understanding of the effects of a high-altitude jet stream on the weather movement and location. The local forecasts were distributed to hotel lobbies, newspapers, radio stations and many public buildings through the local weatherman. So, unless a person just happened to be listening to a radio or made a concerted effort to go somewhere where these forecasts were posted, they would not know the most recent predictions.

People may have read the November 11, 1940 forecast—"cloudy, occasional snow, and colder, much colder"—in the *Minnesota Star Tribune*. However, there was no mention of the terrible wind conditions heading to the unsuspecting Midwest, and most people relied on their own instincts about the changing weather. This was especially true for those who made their living around the Great Lakes. This tragic storm had only one positive result: the National Weather Service went from reporting twelve to fourteen hours a day to an around-the-clock forecasting system. This enabled local weather stations to respond faster to the ever-changing environmental events.

The day before Lake Michigan turned into a freshwater disaster area, the *St. Joseph Herald* reported on the growing conflict in Europe and how Greek troops trapped fifteen thousand Italian soldiers in the mountains and forced them to surrender. Other news included Winston Churchill expressing "good cheer" over the reelection of Franklin Delano Roosevelt. Closer to home, the newspaper reported that the steamer *Sparta* was still aground near Munising, Michigan. (The storm that hit the next day turned the *Sparta* into a total loss.) Also, a druggist in Kalamazoo was shot during a robbery attempt, and the movie *Angels Over Broadway*, starring Douglas Fairbanks and Rita Hayworth, was opening in theaters that night.

The only ominous clue of the terror to be unleashed on the big lake was a small article about how engineers were considering rebuilding the collapsed Tacoma Narrows Bridge. That storm was now bearing down, and no one in the unsuspecting midwestern part of the country could imagine the hellish mix of high wind, cold and snow. Instead, people were enjoying Indian summer conditions and were content to be outside, dressed in light summer clothing.

Before the storm hit Lake Michigan, duck hunters in Wisconsin had gone out that morning dressed for the unseasonably warm weather. When the frozen winds were unleashed, they caught the hunters in their duck blinds and in small boats on the lakes. Many of these hunters either froze to death or drowned when the inland lakes' waves swamped and tipped over their boats. They did not have adequate clothing or protection for the sudden drop in temperature and harsh snowy conditions. The immense violence and suddenness of the winds and snow trapped the hunters in their blinds and boats to suffer slow and agonizing deaths. Those in boats also faced the hellish waves on normally calm inland lakes.

At about 1:30 p.m., five miles off the South Haven pier, the *Richard H.* was sighted laboring in the growing seas. Her futile attempt to reach safety came just before the insanity of the storm grew to an ungodly proportion, powered by the sixty-mile-per-hour screaming winds. The violent wind made the waves roll to incredible heights. After thirty minutes of this torture, a mist covered the lake, blotting out the *Richard H.* and her crew from all earthly eyes for the rest of eternity.

Reports of the *Richard H.* struggling off the South Haven entrance prompted the Coast Guard's commanding officer to order four men to prepare the thirty-six-foot motorized lifeboat to search for the missing fishing tugs. Elmer Dudley, boatswain mate, was in command, and his crew included Machinist Mate Kenneth Courtwright, Surfman Jesse Meeker and Seaman Alvin James. Alvin James had only been in the Coast Guard for a few months and volunteered for the fun of it. Little did he know the chaotic conditions to which he would be subjected. The motorized boat left the safety of South Haven around 2:50 p.m. and rode into the full intensity of the storm. Groups of people, wives, friends and relatives of the Coast Guard men and fishing tug crews had already started a long vigil on the beach. They built bonfires to keep warm, hoping to see the missing fishing tugs. They watched in fear as the lifeboat battered its way out into the mountainous waves. Each time the lifeboat was not visible, the spectators on shore were sure that the crewmen would not survive. Finally, the lifeboat turned north and disappeared, and four more names were added to the prayers said on the beach that horrible day. The lifeboat would not be seen again for the next day and a half. At the Coast Guard station, people nervously waited for the phone to ring. Agonizing hours passed in silent desperation that horrible endless night, hoping for news that their loved ones were alive and in a safe harbor.

In 1940, there were no cellphones or other means of instant communication. The only way news was passed was by radio, telegraph or landlines. Televisions were not yet standard in households. Telegraph and telephones lines were usually the first casualties of a storm, worsening the situation. Furthermore, the early radios were unreliable in storm conditions. Having a radio in small boats like those used today in rescue attempts was considered a luxury and rarely installed. So, for the men in the rescue boat to report their location, they would have to get safely to a shore station.

When the thirty-six-foot lifeboat left the safety of South Haven, it was the beginning of a tense twenty-five-hour trip in thirty-five-foot waves and blizzard conditions, with snow blotting out all land references. The boat averaged only five miles per hour, just a little more than a person, who generally walks about three miles per hour. According to Boatswain Mate Dudley, they went as far north as Holland, about twenty-five miles away, and then turned to go farther out in the lake. Then the lifeboat went south past St. Joseph, adding another sixty-eight miles to their trip. All these miles, they were battling thirty-five-foot waves (almost three stories high) and winds up to sixty miles per hour.

Offshore in St. Joseph, Michigan, Boatswain Mate Dudley determined that the wind and waves were so life-threatening that it would be impossible to enter the St. Joseph Harbor safely. He turned the lifeboat to the southwest to cross the lake and desperately hoped that they could make it into the Chicago Harbor. That hope depended on whether the landmass on the lake's other side could block the winds enough to calm the waves. They had already been out in the freezing, turbulent lake for about eighteen hours. Dudley's crew's spirits dropped when the boat turned away from the dim beckoning lights of St. Joseph and headed southwest toward Chicago. Cold, monstrous gray waves and freezing snow still blinded them. Little did they know when they turned toward Chicago that evidence of their vain search was beginning to wash up on Lake Michigan's shores. Wreckage from both missing fishing tugs were found on the beach north of South Haven, along with life jackets and a few bodies of the crew of the *Indian*.

The Coast Guard officials, friends and neighbors sensed that it would be only a matter of time before clues would wash ashore, telling the fate of the missing, and now presumed lost, lifeboat. The officer who sent these men on what he deemed was a suicide mission must have been going through his own special torment, waiting for any news. Adding to the frustration, no other Coast Guard ships were available to search for the missing men. The only other Coast Guard ship on Lake Michigan was the *Escanaba* (which later sank on convoy duty in the Atlantic with only two survivors), and she was in dry dock in Manitowoc, Wisconsin, for repairs. All other Coast Guard cutters used on the Great Lakes had been sent to the East Coast due to the increasing threat of war in Europe. The only information the commanding officer could give his fellow townsfolk was his personal hope that they were safe in another harbor and unable to send any news of their safety because of downed phone, power and telegraph lines.

After twenty-five grueling hours of adrenaline-infused stress battling waves that towered as tall as three-story buildings, as well as no sleep or food, and all in freezing snowy conditions, the four men finally reached the safety of the Chicago Harbor. Elmer Dudley, hands aching from constantly steering and controlling the throttle, almost collapsed after the boat's ropes were secured to the Coast Guard dock. Only his sense of duty sustained him, and his first thoughts were of the men he commanded. Foremost on his mind was the desire to let his crew members' loved ones back home know that they were alive and safe. As he came into the Chicago station, the officer in charge was in disbelief, as if he was looking at ghosts. The officer had already heard of the four men from South Haven who went out in the killer storm and were presumed lost. The last thing he expected walking into his office was those frozen men.

Elmer Dudley immediately phoned the South Haven Coast Guard station but found that the phone lines were down. He tried to contact them by radio, but the station's radio tower was damaged. He would not be able to get a message out until it was repaired. In the meantime, Dudley ordered his men to get dry clothing and warm food and get some sleep. Finally, almost two days after they left the safety of the Black River in South Haven, the phone lines were back in service, and the men were able to let their home station know that they were safe in Chicago.

The news up to this time was all bad. It only reported wreckage and bodies on the beach and produced a very depressing air around South Haven. The report of the Coast Guard men safely in Chicago hit the people of South Haven like a lightning bolt of good news. The wives of Kenneth Courtwright and Jesse Meeker were finally able to talk to their husbands. The women, red-eyed from lack of sleep, were relieved from the terrible strain of waiting for the fate of their missing husbands. At the same time, they grieved for the families and friends of the fishermen missing on the *Richard H.* and *Indian*.

As the search continued, the magnitude of this freshwater hurricane was apparent. Bodies from the *Anna C. Minch* and the *William B. Davock* were found. Also, the remaining crew from the *Novadoc* had been rescued by the fishing tug *Three Brothers*. The *Novadoc* was declared a total loss. The only body recovered from the *Richard H.* was that of thirty-five-year-old John Taylor. He was first misidentified as Stanley White, but his three brothers correctly identified him. John Taylor's watch had stopped at 2:52 p.m., the time he must have entered the water. Most likely, it was also when the *Richard H.* sank and the lifeboat left South Haven. She fought the storm only an hour and a half after being sighted off the South Haven breakwater. The death toll continued to climb, as duck hunters in Wisconsin were found and two people in Chicago were killed when the high winds blew them off a bridge. They drowned an icy death in the Chicago River. As the cost of the storm was updated daily, the four brave Coast Guard men readied their craft for the trip home to South Haven.

Elmer Dudley decided to hug the southern shore of Lake Michigan on the way home. Even though this route was longer, it would be safer since the lake was still in a nasty mood. On Wednesday, November 13, 1940, they left Chicago and headed home. They stopped in Michigan City and slept there before departing for South Haven the next day. Although mourning the loss of the two fishing tugs, the people of South Haven recognized the courage the four Coast Guard men had demonstrated on this "truly heroic mission." Captain Johnson of the South Haven Coast Guard station asked the St. Joseph station to inform him as soon as they saw the boat pass.

The entire town of South Haven planned to greet the crew. A lookout was set up at the power station, ready to blow its whistle five times when the boat neared. That was the signal to the South Havenites to proceed to line the river and breakwater. The townspeople wanted to give the heroes the heartfelt recognition they deserved. When the power station whistle finally blew, about 1,500 residents lined the shore and gave the four men a true hero's welcome home. The South Haven Drum and Bugle Corps played for the incoming lifeboat as it came into the harbor. Then later that night, there was a dinner at the Yacht Club honoring the men.

Of course, many local events in 1940 became minor news compared to the horror the world was suffering as it became entangled in a world war. Hitler's armies were marching across Europe, and the troops of the Empire of Japan were committing unimaginable atrocities against the people of China. The acts of four brave men who risked their lives in a failed attempt to save others were not nationally newsworthy. Most people at the time did not know much about the Great Lakes and viewed them as little more than oversized ponds, a common misconception of those who don't live near these lakes that we call "Great." Therefore, four local men's uncommon courage and bravery were soon forgotten, along with the great Armistice Day Storm of 1940. Nevertheless, both the citizens of South Haven and these heroic men believed in their hearts that they had demonstrated grit.

Grand Slam in Grand Haven

Grand Haven, also known as "Coast Guard City U.S.A.," is a thriving, bustling city on the east coast of Lake Michigan. For a long time, area residents have had a serious love affair with the Coast Guard and a commitment to honor these brave men and women in an annual citywide Coast Guard festival. The maritime history of this wonderful city affirms that Grand Haven was once the busiest port on the Great Lakes. The number of ship arrivals and departures often exceeded ports in cities like Chicago, Milwaukee, Detroit and Cleveland. Grand Haven operated as many as six lumber mills on the shore of the Grand River. This river is the longest river in Michigan, measuring a whopping 252 miles from its source to its southern terminus, where it empties into Lake Michigan. The river provides access for the city of Grand Haven to the interior of Michigan. This allows goods manufactured and grown as far away as Grand Rapids access to markets all over the Midwest. The riverbanks were home to much industry, including shipbuilding, ironworks, a thriving commercial fishing business and manufacturing sites for flour, shingles and steam engines. Its lumber mills worked seven days a week, twenty-four hours a day. In 1883, Grand Haven produced an astounding 198,092,190 board feet of lumber for other cities around the lakes. The lumber was used in building ships, housing and manufacturing in the bustling and growing cities of the Midwest.

The area's unique history began in the early 1670s when explorers like Father Marquette, La Salle, Louis Joliet and Tonty each visited the mouth of the Grand River. In October 1776, the HMS *Felicity*, a small British sloop, entered the mouth of the Grand River and reported that it would make an excellent port for the east coast of Lake Michigan. Around 1783, Joseph LaFramboise started a trading post at Gabagouache, which means "Big Mouth" in the native tongue. As the population around the mouth of the Grand River grew, so did a natural shift in the mode of transportation, from birch-bark canoes to sailing vessels. This activity led to the first significant shipwreck, the *Andrew*, in 1826. She missed the river's entrance in a storm and ended up beached on the shore. Rix Roberson, the owner of twenty barrels of whiskey in the cargo hold of the *Andrew*, salvaged the whiskey and buried it in a nearby sand dune to prevent people from stealing and hauling his cargo away. When he returned later to transport the whiskey to his trading post, the dune's features had changed, and he could not find the place where he had buried the whiskey. Supposedly, it has never been found, so somewhere near the mouth of the Grand River, there are twenty barrels of wonderfully aged whiskey waiting to be discovered.

In April 1871, Congress authorized $200,000 for the secretary of the treasury to start hiring men at coastal stations to render aid to ships and their crews in distress. Even though the U.S. Life-Saving Service still was not an official branch of the government, this action started the wheels turning for its creation. By 1878, Summer Kimball was appointed to create and be in charge of the service and turn it into a professional organization. He removed the political influences that plagued the service before his appointment. Many of these influences were keepers with political connections in charge of stations who did not know the first thing about saving lives during a shipwreck. Summer Kimball eliminated these keepers immediately and made many of the stations operational full time during the shipping season. He put responsible, professional men in charge and got rid of the political hacks who had afflicted the entire organization.

Summer Kimball established regulations and incorporated a required weekly training schedule. By 1879, many lifesaving stations were manned by full-time crews and classified as first-class stations. However, some stations, like the one in Grand Haven, were manned by only a part-time crew and ranked as second-class stations. Instead of getting a monthly wage, the part-time surfmen were paid ten dollars when they performed an actual rescue and saved a life and three dollars for attending a training session. This policy changed in 1878 to pay ten dollars each time assistance was needed, regardless of whether a life was saved or not. The training standards in both First- and Second-Class stations increased the professionalism of each lifesaving station. The results were immediately reflected in the annual report of the service.

Stations began having weekly training of their men, which was reflected in the number of lives and cargo saved. The respect and reputation of the U.S. Life-Saving Service began to grow in the public's eyes. The men became local heroes just by being in the service and were known as "Storm Warriors." The public often came to watch the men training on the beach and in the water. The most popular of these daily exercises to view was setting up a simulated mast of a ship and using the Lyle gun to fire a projectile over the mast and rig lines so they could rescue the survivors from the ship. Often, ladies from the crowd, including some of the men's girlfriends or wives, were asked to volunteer to be a survivor. The boat righting training was another popular exercise. The lifesaving crew would take the lifeboat or surfboat out into the water and intentionally capsize it. The crew and keeper would then right it and climb back in. If the team was good enough, the keeper, who was in the back of the boat, would not get wet. The crowds on the shore would loudly cheer for their heroes. Kimball also instituted a training manual into the service that required a specific drill for each day of the week, except for Sunday.

The training was scheduled, but since Station 9 was not a First-Class station, not everyone was required to participate. These men only needed to show up when they could or when they were able to make it to a rescue. They were not required to perform beach patrols or man the watchtower every day during the shipping season, as required at a full-time station.

The Grand Haven station was an exception. When the weather got bad, the men would organize daily and evening patrols. The men initiated these patrols and did not receive any compensation. They would start patrolling the beaches north and south of the station and would go out on the breakwater to check for ships in distress. Anyone who has stood on a Lake Michigan beach in November knows how it feels when the sky is dark and angry, raining, spitting snow, and the waves are rolling monsters intent on wiping a person off a beach.

As the number of ships using the Grand River increased, the city of Grand Haven was plotted, and the population grew. Lighthouses, breakwaters, dredging and other harbor improvements were implemented to accommodate the increased volume of ships using the harbor. On December 9, 1860, a significant shipwreck occurred—not at Grand Haven, but at Port Sheldon, south of Grand Haven. The schooner *Vermont* with Captain Albee in command missed the entrance of the port in a storm and grounded some distance off the beach. Strong gale-force winds and waves pounded the schooner to pieces. In a desperate act of survival, the crew started going up the masts to escape the cold waves. The *Vermont*'s first mate, Richard Connell, decided that the crew's best chances lay in someone getting to shore. So, he tied a line around himself and dove into the freezing waves. Struggling against exposure and drowning, he swam the entire distance to the beach. Barely dragging himself out of the surf, his survival instincts and determination to save his crew powered him to rig the line that he had tied to himself. The rest the *Vermont* crew were brought safely ashore. Eleven years later, Richard Connell became the first keeper of the Grand Haven Life-Saving Station.

Initially, the U.S. Life-Saving Service was organized into twelve districts, and three of those districts encompassed the Great Lakes:

- Ninth District, headquartered in Buffalo, covering Lake Ontario and Lake Erie
- Tenth District, headquartered in Detroit, covering Lake Huron and Lake Superior
- Eleventh District, headquartered in Grand Haven, covering Lake Michigan

The Eleventh District's first superintendent was Captain William R. Loutit, who had hired Richard Connell as the keeper of Station 9 at Grand Haven. Captain Connell was responsible for training and instilling dedication into the crewmen he selected. This dedication led to the extraordinary rescues of November 1, 1878.

Actually, these remarkable rescues began almost one month before. On October 11, a howling gale started out as a gentle breeze and was blowing on Lake Michigan since the late afternoon on October 10. Struggling in the dark the entire night, the schooner *Alice M. Bears* of Chicago, Illinois, was trying to make it into the safety and calm waters of Grand Haven Harbor. She was a wooden two-mast schooner, and at 105 feet long and 25 feet wide, she could haul 147 net tons of cargo. Built in 1864, she was in her prime for a freshwater vessel.

The exhausted captain spent the night being beat by frigid waves and wind and could hardly move from his post at the helm. Lining up the *Alice M. Bears* to the entrance of the harbor, her captain fought fatigue, and the schooner barely missed the entrance at 4:30 a.m. He drove her until she rested on the sandy bottom about fifty yards north of the breakwaters.

Keeper Connell and his crew quickly responded when she hit the sandbar. Launching the surfboat through the stormy breakers, they pulled up on the lee side of the *Bears*. The well-trained lifesavers pulled seven exhausted crew members off the wreck. Bundling them as well as possible in the bottom of the surfboat, the lifesavers rowed for shore, backing the surfboat efficiently through the breakers and landing safely on the shore. When the crew of the *Alice M. Bears* was secure, the lifesavers returned to the wreck and scuttled her so she could rest easily on the bottom. In shallow water, she could be easily salvaged later.

Later that same day, the barge *C.O.D.* of Grand Haven sprang a leak in the twisting, pounding waves. The 140-foot-long schooner-barge with three masts could carry 274 net tons of cargo. Built in Grand Haven in 1873, she was a familiar sight in the harbor, hauling mainly lumber from the mills that lined the shore in Grand Haven. The lifesavers observed the *C.O.D.* settling low in the water and set out in their lifeboat to board her. Once on board, they began tossing off her load of lumber to lighten her. The lake would wash this lumber up onto the beach, where it could be easily recovered. The lifesavers helped her crew pump her out, secured her to a towline from a tug and maneuvered her safely into the harbor.

Later that stormy month, on October 28, the schooner *Presto* of Grand Haven ran aground just north of the piers. At 111 feet and a width of 25 feet, she could carry 174 net tons of cargo, and fully loaded, she had a depth of 10 feet. Fighting another raging gale, the *Presto*'s captain misjudged her approach into the entrance, like most accidents at Grand Haven. Not wasting a second in their quick response to the wreck, Captain Connell and his crew were successful in taking off another seven crew members and one passenger. Again, seeing the *Presto* pounded into the bottom by the waves, they re-boarded her and scuttled her in shallow water. This action was to limit any further damage and strain to the ship, a common practice to make the salvage effort easier when the lake was in a calmer and better mood.

As the lifesaving crew recovered from the battle with the lake and discussed the day's events, the watch in the tower observed that the schooner *Persia* was way off course. She finally stranded south of the piers. The *Persia* was a small two-masted schooner with an overall length of only ninety-six feet and a width of twenty-one feet. Having a full load of ninety-two net tons, she only drew seven feet of water. Usually, the *Persia* could get over most of the sandbars and ground close to shore. In total disbelief at the prospect of rescuing another shipwreck, the crew responded quickly by launching the surfboat. When the lifesavers got to the scene of the stranding, they found the *Persia*'s crew already safe on the shore. The crew, feeling the crunch of the sand when she ran aground, had launched the *Persia*'s only lifeboat and successfully navigated it through the treacherous waves to land on the beach. The lifesaving crew made sure that the crew members were in good shape and invited them to the station for dry clothes and food.

Typically, the crew of a merchant vessel is not trained to navigate small boats through the breakers along a shore. This action by a merchant crew usually ends in disaster, with some or all the crew drowning. Using the lifeboat of a sailing schooner is intended as a last resort. Launching a small boat from a sinking ship in a storm is often a frantic act of desperation, but launching it to go through the wild waves after hitting a beach is an act of foolishness. So, it is likely that there were many comments from the lifesavers about the good luck and foolishness of the *Persia*'s crew.

The storm that beached the *Persia* continued until October 29, and its next victim was the schooner *George W. Wescott*. Not a lot is known about the *Wescott*, except that she was from Kenosha, Wisconsin, and ended up on a sandbar just north of the piers. Keeper Connell and his crew responded with their usual efficiency and rescued seven men. The lifesaving crew went to work pumping her out by hand and removing her from the sandbar. Thankfully, they were able to bring her safely into the harbor.

Ships were wrecking at Grand Haven at an astounding rate. Later that same night, the *H.B. Moore*, a 120-foot, 184-net-ton schooner stranded half a mile from the piers. With the storm still raging and creating a violent surf, the crew of eight wasted no time attempting to reach the beach by abandoning the *Moore* and foolishly trying to row to shore in the ship's lifeboat. The events happened so quickly that the lifesaving crew could not launch their lifeboat or get lines out to the *Moore*. Instead, they bravely waded out into the violent, freezing surf to retrieve the men from the *Moore*. It was a long, cold, wet walk back to the station for both the lifesaving crew and the *Moore*'s crew. Once back at the station, the lifesavers gave the men dry clothes, warm food and coffee—perhaps a bit of alcohol was mixed into the coffee.

October was a dramatic prelude to what Lake Michigan was brewing for the lifesavers at Grand Haven. Saving at least thirty-four lives in one month was an astounding feat, but that was only a warmup to the events that would grand-slam them on the first day of November. It is common for the rate of shipwrecks on the Great Lakes to go way up in November. Following is a breakdown of the percentage of shipwrecks per month:

Shipwrecks per Month

Month	*Percentage*
January	0.74
February	0.20
March	1.75
April	5.74
May	7.29
June	4.72
July	6.01
August	7.15
September	13.77
October	21.46
November	24.83
December	6.43

The lifesaving crew only had two days of rest between Tuesday, October 29, and Friday, November 1, 1878, before they demonstrated how remarkably they could perform on the Great Lakes during the cruel month of November. Late on Thursday, October 31, the wind shifted and started to blow hard, turning into another vicious gale. A mixture of snow, rain and a freezing chill accompanied the fierce winds. As the impenetrable darkness turned to a cold, gray, gloomy morning, the winds were now blowing a full gale. Waves were splashing violently over both breakwaters, and Keeper Connell and his crew knew that it would be a busy day. At 11:00 a.m., Keeper Connell received a telegram from Lighthouse Keeper William Robinson in White Lake, forty-two miles north of Grand Haven. A ship had run aground earlier that morning, and her crew had climbed into the rigging to escape the waves. Local citizens had tried three times to launch a yawl boat from the schooner *Ellen Ellinwood* into the breakers on the beach but failed. Each time, the vicious waves knocked them and the boat back onto the shore. After the third attempt, the locals decided that they needed professional help and telegraphed the closest lifesaving station at Grand Haven.

Today, forty-two miles does not seem that far, but back in 1878, the only roads connecting the two communities were dirt roads used by horses and wagons. The other option was to use the railroad. This situation created a dilemma for Keeper Connell: to take his entire crew or only a portion. Considering the events of the prior month, Keeper Connell decided to take only four men, leaving the other four at Grand Haven with the number one surfman, John DeYoung, in charge. Typically, a lifesaving station included the keeper and eight men, ranked one through eight, with the eighth man being the least skilled and least experienced lifesaver. Many times, when the keeper was absent, the number one surfman would be in charge and carry out the keeper's duties. When an opening at another station or the keeper position at a current station became available, the number one surfman would usually be promoted to fill the vacancy. The lifesaving crew in Grand Haven at the time was as follows:

U.S. Lifesaving Crew, Grand Haven

Name	*Rank*
Captain Richard Connell	Keeper
John DeYoung	No. 1 Surfman
C. Haffenback	No. 2 Surfman
John Fisher	No. 3 Surfman
Cornelius VerMullen	No. 4 Surfman
James Barlow	No. 5 Surfman
R. Carmel	No. 6 Surfman
J.J. Glasson	No. 7 Surfman
Unknown	No. 8 Surfman

The only known ranking is that of John DeYoung. Many of these men were fishermen from Grand Haven and had years of experience in dealing with Lake Michigan, and the prime example was John DeYoung himself. It is not known who went with Keeper Connell to White Lake and who stayed in Grand Haven.

As the men prepared the necessary equipment for the rescue attempt at White Lake, the tower watchman reported that the *Australia*, a 109-foot, 151-net-ton schooner, had hit the north pier. DeYoung and the other three men immediately gathered equipment to assist. With loud crunching and splintering of wood, the boat repeatedly bumped its way down the breakwater. As she kept slamming the breakwater, two of her crewmen panicked and decided that their best chance of survival was to jump from the *Australia* onto the north breakwater. Unfortunately, P.H. Cassady of Salem, Massachusetts, slipped on the icy deck as he attempted to jump. Waves immediately washed him overboard, and he drowned an agonizingly painful death in the cold lake, only feet from safety and the north breakwater. The *Australia*, still under full sail, eventually hit the breakwater hard enough to cause a loud *bang*. With that fatal collision, her bow caved in, allowing tons of cold Lake Michigan water to pour in, and she continued hammering her way down the breakwater. Finally coming to a resting place, the four lifesavers heaved a line to her via a throwing stick and secured her to the breakwater. This line prevented her from swinging broadside to the beach and allowing wave action to continue to damage her. The lifesavers rescued the six remaining crew from the *Australia* onto the wet and icy breakwater by noon. Along with the crewman who had safely leaped the breakwater earlier, they made their way to the station for dry clothes, hot coffee and food. After the storm, P.H. Cassady was never found, his body forever claimed by Lake Michigan.

Keeper Connell, by this time, had loaded the Lifecar and the beach cart onto the special train for White Lake. The Lifecar is a specially designed piece of lifesaving gear. Created by Joseph Francis, it was a completely sealed boat with a hatch that opened so eight men could enter and close the hatch behind them. Once securely inside, the Lifecar could be pulled through the waves along a guide rope to shore. It is basically a variation of the breeches' buoy, but now eight could be pulled to safety instead of one man at a time. Once the equipment was on the train, Keeper Connell was informed that the shipwreck at White Lake was the *L.C. Woodruff*, a 170-by-33-foot, three-mast bark that could carry 549 gross tons of cargo.

On October 31, 1878, the *Woodruff* anchored off White Lake, and during the night, the northwest gale increased in violence, blowing out most of her sails. With only her mainsail, she was practically powerless. Around midnight, the winds switched from the northwest to the southwest, causing the big bark to drag her anchors. By 4:00 a.m. on November 1, her stern hit the first line of breakers near the beach. Around 8:00 a.m., she hit a sandbar, effectively grounding her in thirteen feet of water. The ship took a severe beating, and her seams opened. Water rushed into her hull, finally settling her on the bottom. As the violence of the southwest gale continued to sweep waves over her deck, the crew's only hope for survival was to climb high up the ship's masts. With a loud *crack*, the mizzenmast went over the side. With the furious seas sweeping the deck, the crewmen knew that their only chance of survival would be climbing high enough on the remaining two masts. But there was still only a slim chance for survival, as they had settled about half a mile north of the piers at White Lake entrance, right in front of the local sawmill.

Knowing the size of the *Woodruff*, Keeper Connell reasoned that she would draw more than ten feet of water and guessed she would be some distance from shore. It took two hours to travel the forty-two miles by train, and the assistant lighthouse keeper, Thomas Robinson—William Robinson's eighteen-year-old son—met them there. He had the lifesavers transfer all the equipment to a tugboat. The tug took the lifesavers on the inland waterway of White Lake to the point closest to the wreck site. They unloaded all the equipment with the help of the local citizens. There were about one hundred yards of beach, so it was not the ideal spot for the rescue. The area was clogged with the refuse of sawdust, slabs and edgings from the local sawmill. This debris formed a twelve-foot-high mound and would be a major obstacle in shooting a line out to the shipwreck. The *L.C. Woodruff* was located in thirteen feet of water about 150 yards out from the shore. Her mizzenmast had already gone over the side, and her ten-man crew was hanging on the other two masts. Making matters worse was another submerged shipwreck between the edge of the beach and the *Woodruff*, creating another obstacle. Nevertheless, Keeper Connell and his four men hurriedly set up the Lyle gun to shoot a line out to the *Woodruff*.

Local citizens, who were overly eager to help, hindered the lifesavers' rescue efforts. On a sandbank, behind the spot where the lifesavers were setting up their equipment, a large crowd of townspeople gathered. The group included crews from ten to twelve vessels that were sheltering in White Lake from the storm. Many of these men were captains and crews of the schooners in port. They had participated in the failed attempts to rescue the crew of the *Woodruff* earlier that day, and they were anxious to assist now. Not only did Keeper Connell contend with a twelve-foot mound of debris and a submerged wreck that could tangle his line, but the overzealous crowd also kept getting in his way.

With the professionalism of a highly trained crew, the keeper set up his Lyle gun, and the first shot sent a line over the *Woodruff*, just aft of the main rigging. With the line now on the shipwreck, the *Woodruff*'s crew started pulling the one-and-a-half-inch line and the tail block out to the wreck. About forty yards out, the monstrous seas and a strong current running parallel to the beach entangled the shot line in the submerged wreck. Simultaneously, another part of the line got tangled in the jagged edgings of the debris mound. With the lifesavers trying to clear the line from shore and the crew of the *Woodruff* continuing to haul on the line, the shot line parted and the zealous crowd started hauling in the line. Now the crew of the *Woodruff* had the remains of the shot line, and the crowd had the larger line pulled back on the beach. Keeper Connell had to start over at square one.

The Lyle gun was charged with powder again and a projectile inserted into the barrel. With a resounding roar, the shot line fell short of the *Woodruff*. The cause of the missed shot was likely a wet line that added extra weight. The third attempt used a dry line, and this time, they were successful in carrying it into the rigging, where the crew could retrieve it. The lifesavers had learned from their previous experience and were more cautious feeding the heavier line out to the shipwreck. The tail block reached the men on the *Woodruff*, but the retrieval line to get the tail block back to shore somehow got entangled in the submerged wreck. While trying to free this line, the enthusiastic crowd attempted to take the yawl from the schooner *Ellen Ellinwood* again, guiding it out using the line shot out to the *Woodruff*. Five volunteers got into the boat and started hauling themselves out to the wreck. The waves overturned the boat, and now the five volunteers had to be rescued too. Between controlling the crowd's antics and the difficulties of the wreck's location, it was starting to grow dark. Keeper Connell decided to try to send the Lifecar out to the wreck. His intention was to haul the Lifecar out to the wreck using the suspended whip line. Once it reached the shipwreck and loaded with some of the survivors, it could be easily retrieved.

Three factors—the time wasted with the onlookers' unwanted help, the difficulties with the entangled lines and now the poor visibility of the oncoming darkness—caused the captain of the *Woodruff* and some of her crewmen to act precipitously. Had they waited for the Lifecar, most likely, all the crew would have been saved. Instead, the captain and three men started pulling the *Woodruff*'s lifeboat along the suspended whip line. When the crowd shouted, the lifesavers looked up from their Lifecar preparations and saw a mob of at least fifty men rushing the beach side of the whip line and starting to haul on it. Without heeding the warnings of the lifesavers, the volunteers strained the line to such a degree that it snapped at the *Woodruff* side of the line. The four men in the boat had no choice but to hold on because the crowd was pulling them. The boat submerged as it was dragged to shore. Because the captain was in front of the boat, he was uninjured, but the other three men were beaten senseless by the oscillations of the submerged lifeboat. Once on shore, the lifesavers went to work to revive them. Although two of the three unconscious men were revived, the *Woodruff*'s first mate died of injuries and drowning.

Now the *Woodruff*'s six remaining crewmen, seeing the only connection to land severed and the wreck breaking up in the growing darkness, hastily made rafts from doors and deck planks. In total desperation, they jumped into the freezing water and attempted to swim for shore. One of the crew members immediately sank, and his body was never found. Another came ashore about a mile and half up the beach and a third two miles up the beach. Two of the men reached the beach close to the debris mound. The last man was never seen leaving the wreck. Out of the ten-man crew, two drowned and one, the first mate, died of his injuries. Seven crew members were rescued and three lost, but even that was considered a miracle because of the adverse conditions Keeper Connell and his lifesavers had encountered. The U.S. Life-Saving Service's investigation absolved Keeper Connell of any blame. The report confirmed that the five lifesavers were totally incapable of taking effective measures to control the large, unruly and overzealous crowd.

Back in Grand Haven by noon, the gale did not show any signs of abating, so surfman John DeYoung had the surfboat taken down to the beach in case it would be needed. Shortly afterward, the 138-foot-long schooner *America* hit a sandbar just north of the piers. Launching the surfboat into the wild waves, surfman DeYoung and his three men reached the grounded *America*. Carrying about 250 tons of iron ore, she was firmly planted on the bottom. The lifesavers skillfully pulled up on the lee side of the schooner and rescued the entire crew of eight men. Turning and heading back to shore, DeYoung showed incredible skill in handling the surfboat back through the breakers and landing all the occupants safely on shore.

Around 3:00 p.m., the schooner *Elvina* of New York missed the entrance and came onto a sandbar between the north pier and the *America*. With her stern swinging around, she hit the bow of the *America*. The smaller schooner was driven close enough to shore so the lifesavers could wade out into the surf and was able to get a line to her, thus preventing further collisions between her and the *America*. At the request of the captain of the *Elvina*, the lifesavers rowed the surfboard out to the wreck and brought the captain safely on shore. The remainder of the *Elvina*'s crew chose to remain on the ship.

After they finished with the *Elvina*, the 138-foot-long schooner *Montpelier* misjudged the entrance at Grand Haven and hit a previous wreck, the steamer *Orion*, which sank in 1870. The impact with the *Orion*'s iron machinery blasted a hole in the wooden schooner, and she filled immediately, sinking close to shore. The lifesavers responded by launching their surfboat for the third time on that wicked afternoon. With the crew of the *Montpelier* in the rigging, they rowed out to the fifth wreck. Battling their way through the vicious surf, they pulled their surfboat up to the lee side of the wreck. There, they got the seven men and the female cook to board the surfboat. With all of the shipwrecked crew bundled in the bottom of the surfboat, DeYoung, with his three surfmen pulling at the oars with frozen, aching muscles, landed all the occupants through raging surf, safe and snug on shore.

The exploits and difficulties these eight men of Grand Haven Life-Saving Station 9 overcame on November 1, 1878, are what makes legends and heroes. This group of volunteers was not seeking the glory that followed this eventful day. Instead, their motivation was a straightforward calling: the duty to assist others in distress, much like today's heroes in the Coast Guard service. Typically, a station performed one, maybe two shipwreck rescues per day, but Station 9 accomplished a grand slam in Grand Haven.

In Search of Jack Dipert's Ghost

This story began as a tale of the shipwreck of the *Henry Cort* at Muskegon, Michigan, and the heroic rescue of her crew by members of the U.S. Coast Guard. There are various resources about this rescue, and it is documented that one person was lost during the first rescue attempt. However, there is very little information about the life and death of U.S. Coast Guard surfman Jack Dipert. At times while writing this book, I felt like I was chasing a ghost.

The act of sacrificing one's life so others may live, especially strangers, is a perplexing conundrum that scholars and psychologists have debated for centuries. What goes through a person's mind when faced with a dangerous challenge that may result in losing one's life? This question probably was not on Jack Dipert's mind as he climbed aboard the station's surfboat on November 30, 1934. His only concern was the crew of the whaleback *Henry Cort* that was impaled on the breakwater outside Muskegon Harbor. Being outside the harbor exposed the *Cort* to the full fury of Lake Michigan and put her crew in extreme peril. As Jack Dipert climbed aboard the lifeboat, his sole preoccupation was to rescue the *Cort*'s crew. Little did he know on that cold, windy November night that he would become the only casualty. Some say that the lifeboat into which he climbed was a jinxed craft. Whatever his rationale, Jack Dipert, a member of the Coast Guard for only five months, was following in his father's footsteps, having grown up hearing the Coast Guard motto, *Semper paratus*, "Always Ready." His father, William, was a twenty-seven-year veteran of the Coast Guard, and at the time of his son's death, he was in command of the Point Betsie Station just north of Muskegon.

To understand the loss of the *Henry Cort*, one must understand her design and characteristics. The *Henry Cort*, originally named the *Pillsbury*, was a design called a "whaleback." She and forty-two of her sister steamers and barges were common sights on the Great Lakes. Although a few of these ships sailed the oceans, the majority were kept in the Great Lakes. Because of her unique design, sailors on the Great Lakes commonly called this type a ship a "pigboat." She had a long, narrow hull with rounded sides and a flat bottom, giving the ship a very low profile. All the elements of her hull terminated at the bow with a flat nose that resembled a pig's nose. When loaded, these ships' beauty was certainly in the eye of the beholder.

Captain Alexander McDougall of Duluth, Minnesota, came up with the idea of the whaleback while fighting a storm in a conventional steamship on the Great Lakes. His theory was that the ship would offer less resistance to the wind if the sides were rounded instead of flat. Furthermore, if it sat low enough in the water instead of above, the weight of the waves would be welcome aboard because that weight would help steady the ship. With rounded turrets above the main deck and most of the ship below water, the resistance to the wind would be minimal, and its stability would be maximized. McDougall also promoted the idea of making ships out of steel instead of wood. In the 1880s, steel construction was not a standard practice anywhere. He figured with steel, he could make the ships bigger and stronger and able to carry more cargo. Further, the cost of building such a ship would be far cheaper than with a wooden vessel. Captain McDougall applied for patents and created models of his ship design so he could show potential investors. Unfortunately, the model ships did not look like anything currently in use in the shipping industry. His potential investors ridiculed him with negative comments about the design being a high-risk financial investment.

Finally, in 1888, using his own money, he built his first whaleback. It was an unpowered barge that would have to be towed by another ship. At that time, many types of barges were common on the Great Lakes and were considered an economical method of transporting cargo.

A barge required less maintenance and fewer crew members. A powered vessel and the barge could be loaded simultaneously, thus reducing cost and time. It could also carry more cargo than a single powered vessel. Barge 101 was launched on June 23, 1888, in Duluth, Minnesota. Loaded with the rich iron ore from the Minnesota ranges, she delivered her cargo to Cleveland successfully. Once she had proved its worth, Captain McDougall was able to

Henry Cort leaving Duluth. *Alpena County George N. Fletcher Public Library, Great Lakes Maritime Collection.*

gain financial backing from John D. Rockefeller to build more of his designs. Eventually, forty-three of his barges and steamships were built. The only surviving example of this unique design is the museum ship SS *Meteor*. She is on display in the city that built her, Superior, Wisconsin. Exploring her is like stepping into a time capsule.

Hull 125, later named the *Henry Cort*, slid into the water as the *Pillsbury* on June 25, 1892. The *Henry Cort* had an overall length of 335 feet and a beam of 42 feet. She was registered to carry 2,234.49 gross tons and was powered by a 1,400-horsepower triple expansion steam engine. Two Scotch boilers provided the steam for her engine. Her top speed, when fully loaded, registered as fifteen knots.

The *Pillsbury* was built for the Minneapolis, St. Paul & Buffalo Steamship Company and used as a package freighter to haul train cars of freight for the Soo Line Railroad. She made her first trip hauling this type of freight in August 1892. Then she was sold to the Bessemer Steamship Company on June 16, 1896, almost four years from the day she was launched. She was renamed the *Henry Cort*. The year 1896 was a time of great upheaval in the financial circles around the Great Lakes. John D. Rockefeller was buying

The *Pillsbury*. *Alpena County George N. Fletcher Public Library, Great Lakes Maritime Collection.*

smaller companies and consolidating them into one large monopoly. In 1901, U.S. Steel bought and absorbed the Bessemer Steamship Company into the Pittsburgh Steamship Company, a division of U.S. Steel. From 1901 to 1917, the *Henry Cort* hauled iron ore from the rich mines in the Iron Range of Minnesota to the steel mills in the southern parts of the lakes. On her return trips, she hauled coal to the railroads used in the Iron Range mines.

One of the unique characteristics of the whaleback class of ships is the graceful curve of their bow. This shape enabled a whaleback to charge up onto the ice; by the sheer force of her weight, the ice would break. This characteristic naturally made the whalebacks excellent ice-breaking ships, and the *Henry Cort* was used intensively every early spring and late fall. In that capacity, she was one of the many ships that helped extend the navigation season on the Great Lakes. In 1917, when World War I was ramping up in Europe and steel sales were skyrocketing, the shipping season was extended beyond the normal end date, and the *Henry Cort* was pressed into her familiar ice-breaking duties. With an early heavy freeze that fall, the *Henry Cort* was ordered to start breaking ice in western Lake Erie to keep the shipping lanes open as long as possible. With temperatures in the late fall already in the negative twenties, ice had formed in the St. Clair and Detroit river and lake, blocking many of the ships that needed to make the passage from Lake Erie to Lake Huron. Even the big railroad ferries were pressed into the service of breaking the ice. The situation was so dire that the U.S. government assumed the total cost of the ice-breaking efforts for the first time in history. The ship traffic, in both up and down-bound lanes, was at a standstill. The chances of these ships being icebound for the winter was very real and would result in ruinous losses for their owners. Into this desperate climate, the *Henry Cort* and her sister whaleback the *Neilson* were ordered to Bar Point in Lake Erie at the entrance to the Detroit River. Ice had windowed there and blocked the passage in and out of the Detroit River, creating a huge traffic jam.

The ship owners hoped that the *Henry Cort* and *Neilson*, working together, could effectively open a channel and keep it clear and open long enough for their ships to proceed onto their final designations. The *Henry Cort* would slide onto the ice as far as she could, break it and then back up and try to go forward again and break more ice. As she was charging forward, traffic-jammed ships would pass behind her to use the channel she had created. On one of these attempts, she slid forward more than she anticipated and could not easily budge. More power was ordered from the bridge, and with an intense jerk, she slid off the ice right into the path of the *Midvale*. The *Midvale* ordered emergency reverse, but her forward momentum crashed her into the side of the *Henry Cort*, mortally wounding her. The *Henry Cort*'s chief engineer, feeling the impact from the *Midvale* and hearing the noise of all the bells and whistles, ordered the steam dumped from her boilers to prevent an explosion when the icy water from Lake Erie found her boilers.

As the *Henry Cort* settled to the bottom and her crew was rescued, the captain knew that she could not be salvaged until spring. All winter of 1917 and spring of 1918, she lay on the bottom, waiting to be raised and returned to service. Her location was carefully marked so she would not be hard to find in the spring. The following April, when the salvage operation got underway, the Pittsburg Steamship Salvage Team discovered that the *Henry Cort* had moved. She was more than two miles from the location of her sinking and within eight hundred yards of the down-bound shipping lanes. This difficulty was one of many the salvage operation encountered trying to raise her. The team dove down to her many times, plugging holes and trying to bolster her watertight integrity. By June of that year, she was deemed sufficiently watertight and ready to be pumped out and brought to the surface. As soon as she showed signs of movement to the surface, her deck collapsed, causing her to settle back onto the bottom. The only option left to the salvage team was to build an expensive cofferdam around her and use that method to raise her to the surface. Finally, in late September, she saw daylight again and was towed immediately to a shipyard for repairs. Unfortunately, the yards were already filled to capacity because of World War I, and she had to wait until November 1918 for work to begin on her. With her main deck already collapsed, it was decided to modernize

Henry Cort, damaged, 1927. *Alpena County George N. Fletcher Public Library, Great Lakes Maritime Collection.*

her by installing a new deck over four feet higher than her original design. Additionally, the repairs included removing the turrets on her aft section and adding more conventional and roomy crew accommodations.

The *Henry Cort* served the Pittsburg Steamship Line with only minor problems, most of them due to her ice-breaking duties, until the spring of 1927. Then she stranded on Colchester Reef in Lake Erie, almost at the same position where the *Midvale* had struck and sunk her. By then, the whaleback days were numbered for carrying bulk cargos. Bigger, stronger and faster ships were already built that were more efficient than the *Henry Cort*. While *Henry Cort* was on the reef, the Pittsburg Steamship Line abandoned her to the insurance underwriters, and they sold her to the Lake Ports Shipping and Navigation Company. It pulled her off the reef and sent her to a shipyard to repair the bottom damage. It also added two large whirly cranes and expanded her forward deckhouse for the extra crewmen needed for the cranes. These cranes were used to load and unload scrap and pig iron into her holds. These new modifications extended the *Henry Cort*'s life to that of a tramp ship. Her primary duties would involve transporting scrap and pig iron, and she would haul any cargo her owners saw fit, such as slag, coal, stone or sand.

The deck modifications made her handling capabilities poor, especially when she did not have heavy cargo in her holds. When she was not sitting low in the water with the flat surfaces above the waterline, the surface area allowed the wind to push her around. This was not what Captain McDougall had envisioned for his design. This condition could be the reason she had a tougher time keeping out of trouble. In August 1928, she again ran aground on Colchester Reef in Lake Erie. It took more than six days for the wreckers to pull her off the reef, but she finally came off after 2,300 tons of pig iron cargo was offloaded into barges next to her. Again she touched bottom while down-bound in the Livingston Channel of the Detroit River in the spring of 1933. Tugs escorted her back to the Nicholson Dock in Detroit, but she sank just before reaching it. Once again, she spent most of the winter on the bottom until she could be raised and repaired for the next season.

The *Henry Cort*'s last trip began November 28, 1934, at 9:00 a.m., when she left Holland, Michigan, bound across the lake to Chicago. At Chicago, she was to pick a cargo of pig iron for the furnaces at the Campbell, Wyant and Cannon Foundry in South Haven. As she left the safety of Holland, Lake Michigan began to show the effects of a southwesterly gale. The whaleback's progress was slow, heading into the wind. These boats were never very well-behaved in a headwind at sea. While going into the wind and waves, the smooth spoon-shaped bow made it difficult to stay on a

Henry Cort's whirly cranes. *Alpena County George N. Fletcher Public Library, Great Lakes Maritime Collection.*

steady course, especially without cargo that would protect her upper works from the oncoming wind. The modifications of the oversize squared crew quarters and raised decks that increased her wind resistance, above the waterline, compounded the *Henry Cort*'s problems. So did the added square stern housing for extra crewmen and the two big whirly cranes. Gone were the rounded turrets, which would have helped to reduce the topside wind resistance. In sum, the *Henry Cort*'s original design may have given her a better chance in the conditions on Lake Michigan that day.

As the whaleback crossed the lake and struggled to reach Chicago, Captain Charles Cox had difficulty keeping the *Henry Cort* pointed on course into the wind. With each hour that passed, he wished more and more that he were taking the wind and waves on his stern quarter instead of his bow. Twenty-eight miles from Chicago, the gale granted his wish and blew her completely around, pointing her 180 degrees from her original course and back toward Holland, Michigan. Now Captain Cox's choice was to re-cross Lake Michigan with a following sea on her stern. He knew that the western entrance to Holland would be dangerous in a southwesterly gale, so he decided to try for Muskegon, Michigan. At Holland, his course would take the wind and waves broadsides, whereas at Muskegon, he could keep the wind and waves more on her stern as he approached the harbor. The conditions on Lake Michigan that day made both options dangerous, but trying to get into Muskegon was the lesser of the two evils.

As she crept closer to Muskegon, just when Captain Cox lined his ship between the breakwaters and safely into the harbor, the current coming out of the river conspired with the wind and waves on Lake Michigan to create a tremendous sea and lift the *Henry Cort*. To keep her aligned with the channel, Captain Cox ordered the stern anchor to be dropped. It was too little, too late—the seas spun her around like an enormous bobber and smashed her on the north breakwater just outside the harbor entrance.

The scream of her twisting and buckling port side plates opened her hull to the lake and quickly settled her on the bottom. Taking on a thirty-degree list to starboard, she rested firmly on the bottom and riprap stone of the breakwater. Lodged on the bottom, she was about three thousand feet from the safety of the shore and leaning away from the breakwater. Waves relentlessly washed over the breakwater and onto her decks and lashed out at her cabins. Electric lights were the first casualty, leaving her nothing but a cold, dark wreck. As she tilted more and more, dishes stacked neatly in the racks and cabinets of her galley showered the cook, Harry Sutton. Also, the big stove in the galley was torn from its mountings and its smokestack dislodged. With the catastrophic flooding of her hold and engine room, her men were powerless to prevent her from sinking again. As the initial shock wore off, the crew's next fear was that her boilers would explode when the icy water of Lake Michigan hit them. Most of her crew, braving the open storm-lashed deck, rushed forward to the deckhouse to get clear of a potential explosion from her engine room. Fortunately, Chief Engineer Augie Britz mitigated the danger of a boiler explosion. As he abandoned his post to the cold water flooding into the engine room, he had the presence of mind to take steps to prevent a catastrophic boiler explosion.

When the crew realized that the boiler was not going to explode, they went back toward the stern of the *Henry Cort*. There they congregated in the social center of the ship, the galley. Several men fumbled around in the dark and were able to find a lamp and some candles. The light enabled the men to take stock of their situation, and they figured out that it would be suicide to try to make it to shore. Even though they were without heat or electricity, they were safe from the storm lashing outside. The next line of business was to use some muscle and get the big coal stove back in place and her smokestack put back together. With this done, at least they had heat while they waited for a rescue that probably wouldn't happen until morning. Cook Sutton brewed up some fresh coffee and pulled leftover Thanksgiving turkey out for the men to eat. With the immediate danger passed and some food in their stomachs, their situation did not look so bleak. Some of the men played cards, while others went to their rooms to sack out until morning. They didn't realize that the wrecking of the *Henry Cort* mobilized a rescue effort.

On that fateful night, Jack Dipert reported for the evening watch after enjoying the entire day on liberty. Like most young men, he was thinking about a championship football game the next day. He was sent up to the drafty watchtower and spent his watch shivering from the frigid blast coming off Lake Michigan, seeing only a scant distance toward the harbor entrance and the raging waves inside the breakwaters. This was a sign that Lake Michigan was in a furious and violent mood. Apprehension and dread crept in as he stared into the cold, black night. At about 10:00 p.m., the alarm sounded in the station, jolting young Jack Dipert into action. The hours of training took over, and there was no time to ponder the dread that he felt. There was no time for fear or even thinking; instead, the Coast Guard's training kicked in, and he went through the motions that were now second nature. Chief Boatswain Mate John Basch wasted little time alerting his crew. As he gathered the four members—Edward Beckman, Roger Stearman, Charles Bontekoe and Jack Dipert—they hurriedly raced to launch the station's largest power surfboat. Even though the powered surfboat was a stout thirty-six feet long, it was driven by an underpowered forty-horsepower gasoline engine.

All the men knew that they were going out into a perilous situation. As the surfboat slid down the rails into the dark night, it started to feel the waves that had already battered a 335-foot ship to pieces. The closer they got to the harbor entrance, the bigger and sharper the waves became as they blasted through from Lake Michigan. Passing through the entrance and out into the turbulent Lake Michigan, the waves became violent enough that they would launch the boat entirely out of the water. A lesser craft cresting each wave could not have stood up to the punishment. The main problem for Chief Boatswain Mate Basch was the darkness of the night, as black as the proverbial coal mine. He had trouble judging or seeing the height or angle of the smashing waves. Chief Boatswain Mate Basch had only forty pitiful horsepower available, and it was almost impossible to make headway and maneuver the big surfboat. With the relentless waves, the men could only hang on and pray that they would make it past the next wave.

Then the inevitable happened: a series of waves larger and sharper than before smacked the boat head on. Instead of rising and going over them, the boat pushed through the icy waves. When the boat emerged from the other side, Charles Bontekoe, Roger Stearman and young Jack Dipert were not in it. Both Charles Bontekoe and Roger Stearman were able to re-board the surfboat, but Chief Boatswain Mate Basch only caught a glimpse of Jack in the water before the cold black of the night swallowed him up. Now the surfboat's mission changed from rescue to survival, as another series of waves hit the boat, swamping the engine and causing it to lose power. As they approached the shore north of the breakwater, Edward Beckman, with a line tied around him, jumped into the breakers and headed for shore. Once there, Edward pulled Chief Boatswain Mate Basch through the breakers and onto the shore. Then the two men pulled the other two from the surfboat. Local citizens Harold Smedley, George Swarthout and L.H. Powrie responded to the cry of a shipwreck in town and went down to the beach. They ran into the Coast Guard men as they emerged from the breakers and helped pull them out of the freezing water. Exhausted, the four Coast Guardsmen's first thought was to find Jack, hopefully alive. They patrolled the beach for more than two hours, and all they accomplished was finding the lifeboat. There was no sign anywhere of Jack Dipert, the twenty-three-year-old veteran of five months. He had been swallowed by the cold, black, unforgiving lake and gave his life to make sure others lived.

After the futile search for Jack, the men from the surfboat went by ambulance to the Coast Guard station for some food and a much-needed rest. After Chief Boatswain Mate Basch informed his superiors of the loss, they responded by getting more help from the White Lake and Grand Haven Coast Guard Stations. Both sent their largest surfboats to Muskegon. The six-man crew from Grand Haven left immediately in the big lifeboat. Chief Boatswain Mate William Preston accomplished the hazardous twelve-mile trip from Grand Haven to Muskegon in the worst possible lake conditions. As Preston made the entrance to the Muskegon Harbor, the surfboat passed within one hundred feet of the *Henry Cort*. The men noticed lights coming from the ship, an indication that there was still life aboard. Chief Boatswain Mate Preston knew that any attempt at a rescue could not be accomplished until daylight. Also, the powerful Coast Guard cutter *Escanaba* arrived on the scene and radioed that they had turned their powerful searchlight on the *Henry Cort* in the hopes of spotting some life in the ship. All the surfboats rendezvoused in the protected channel of Muskegon with Lieutenant Ward Bennett, the commander of the Tenth Coast Guard District. Bennett decided that he needed to be on site to make effective decisions concerning the rescue. He drove furiously from Grand Haven to Muskegon to access the situation in person. After listening to both Preston's and Basch's reports and the results of the futile search for Jack Dipert, he decided that any other course of action that night would be extremely hazardous. An attempt would probably fail and expose his men to high risk. He had already lost one man and did not want to risk losing more, nor any of the crew from the *Henry Cort*. Bennett decided to wait until daylight to assess the situation and determine the best course of action.

If Lieutenant Bennett and Chief Boatswain Mate Basch had known the actual conditions the men on the *Cort* were experiencing, then Jack Dipert would not have been lost. However, this was a time when communication between a ship and shore only occurred when the boat was tied up at a dock. Very few ships had a radio, and most captains were distrustful of that newfangled technology. They did not want the pencil pushers back at the office second-guessing their decisions and using a radio to "look over their shoulders." Most captains back then, even if they had a radio on board, would not use it, except in an emergency.

The hours before daylight had to be the hardest on Chief Boatswain Mate Basch and the three remaining men of his ill-fated crew. Undoubtedly, Basch was constantly wondering if he could have done something different to prevent the loss of Jack Dipert. He probably kept praying and hoping that the search parties had missed him, or maybe he reached somewhere along the breakwater and would be found in the morning. Losing a person in his command had to weigh heavily on his mind that night. When he saw the boy's father, how would he explain his actions? Worse yet, how would he present the details of the accident to the boy's mother? He knew that he had made the correct decisions based on the available information, but logic and reason are not comforting to a grieving family.

In the cold gray light of morning, Lieutenant Bennett and several men went down to the beach on the north side of the harbor entrance. They saw the condition of the angry lake, its waves crashing against the breakwater and the *Henry Cort*. As the spray landed on top of the rock of the breakwater, it quickly turned to ice. The Coast Guard cutter *Escanaba* had left its Grand Haven station during the night and was stationed off the breakwater, patrolling back and forth. After conferring with the captain of the *Escanaba*, Lieutenant Bennett decided that it was still too risky to send a boat out on the lake to rescue the crew of the *Henry Cort*. The ship was close enough to the breakwater, so they decided that the safest option was to send a team of men roped together to a position opposite the *Henry Cort*. From there, they could alert any surviving crewmen from the *Henry Cort* and rig a breeches' buoy to transfer the men safely to the breakwater. Once on the breakwater, the men could be tied together and transported safely the three thousand feet of the breakwater to shore. It would still be a dangerous rescue, but it was safer than trying to transfer the men to a pitching lifeboat.

A group of Muskegon Coast Guardsmen selected Chief Boatswain Mate Basch to lead out along the north breakwater. Lashed together for their safety, they painfully and carefully climbed over the ice-covered boulders until they were opposite the broken *Henry Cort*. They found Captain Cox and his crew already chopping their way on deck from the ice-coated steel ship. As the cutter *Escanaba* patrolled just offshore as a safety precaution, the Muskegon Coast Guardsmen shot a line over to the *Henry Cort*. After rigging the lines, they were able to transport the entire twenty-five crew members to the breakwater. In the most honored tradition, Captain Cox was the last man to leave his ship after making sure his entire crew was safe on the breakwater.

Once Captain Cox and his crew assembled on the breakwater, the *Henry Cort*'s exhausted crew and the Muskegon Coast Guardsmen lashed themselves together into three groups. This measure was to prevent any of them from slipping into the water. The men had to crawl over icy boulders as the waves from the lake were splashing them. This was the worst part of the entire ordeal

Henry Cort wrecked on the Muskegon breakwater. *Alpena County George N. Fletcher Public Library, Great Lakes Maritime Collection.*

for the crew of the *Henry Cort*. Two of the exhausted men had to be carried: the oldest member, cook Harry Sutton, sixty-eight, and First Mate Harvey Matthews. Harvey Matthews was the second person off the wreck and helped the Coast Guardsmen with the remaining twenty-three men. Once the men reached shore, Harry Sutton was rushed to Hackley Hospital, and Harvey Matthews was rushed to Mercy Hospital. Both recovered from exposure. The remaining survivors from the *Henry Cort* were taken to the Civilian Conservation Corps's barracks and given dry clothes, food and a place to recuperate from their terrible ordeal. Once the weather calmed down, the crew was allowed back on the *Henry Cort* to recover their personal belongings.

After the rescue, Captain Cox blamed the light load in the *Henry Cort* for her sinking because that allowed too much of the vessel exposed to the wind. So, he was unable to steer a steady course when she approached the harbor entrance. He said that a sudden shift in the wind caused her to veer to the north and strike the breakwater as they approached the harbor. After hitting the breakwater the first time and with water rushing in, the wind spun her around and crashed into the breakwater again. This was a fatal blow, breaking her in two. Captain Cox stated that his biggest regret about the whole incident was the loss of Jack Dipert.

Five days after losing Jack Dipert, Lieutenant Bennett conducted an investigation into the death of young surfman, which is required when there is a loss of life during any station or unit operation. In that investigation, Chief Boatswain Mate Basch stated that the lifeboat refused to respond when he threw it hard over to port. Thinking there was a problem with the steering, he threw it hard to starboard, and the lifeboat also "balked in that direction." An examination of the lifeboat, while it was still high on the beach, revealed that there was nothing wrong with the steering. The final report cleared Chief Boatswain Mate Basch of any wrongdoing and concluded that Jack Dipert's death was due to extreme weather conditions. Coast Guardsmen and volunteers of the Civilian Conservation Corps and Sea Scouts patrolled the beaches of Muskegon for months after the shipwreck, looking for Jack Dipert's body. They never found it—it is still out in Lake Michigan somewhere.

On a cold, dark, stormy night in November, one can imagine seeing what appears like a person standing out on the north breakwater in Muskegon near the remains of the *Henry Cort* and wonder what that person is doing there. It may be young Jack Dipert, contemplating why he was left out there alone. But after researching this young man, I prefer to think that he is waiting for another shipwreck and still trying to uphold the Coast Guard motto, *Semper paratus*.

Not a Good Time

Winter around the Great Lakes can be extremely harsh. Each lake, depending on the way the wind blows, is a unique weather generator. Because the weather on the Great Lakes can change so suddenly, typically, ship owners will start laying their ships up for winter maintenance near the end of November. The owners will wait until the ice begins to break up on the lakes before sending them out again. One of the primary reasons for this voluntary layup is that insurance premiums for vessels skyrocket after December 1. Ice is another reason, especially in the northern part of the lakes. Ships stuck in ice in the late 1800s and early 1900s for weeks at a time, sometimes out of sight of land, can run short of food for their crews. There is a story of one such ship, stuck in the ice, where after their food had run out, the men existed on their wheat cargo until they could get free. They ate a wheat mash for the morning, noon and evening meals. Finally, the ice moved the ship close enough to shore, and a party of brave men crossed several miles of open ice to get supplies. The other problem is that ice can do damage to a ship's hull, rudder or propeller.

For these reasons, owners prefer to bring their ships to their home port and tie them up until spring. That way, maintenance can easily be performed, and the crews get some time off. Some owners' practices are exceptions to this rule, especially in the southern part of Lake Michigan in the late 1800s and early 1900s. These owners kept one or two of their boats running through the winter. These boats were generally older, and their loss would not as severely affect the company finances. In the late 1800s, these boats

City of Duluth in harbor. *Alpena County George N. Fletcher Public Library, Great Lakes Maritime Collection.*

typically had wooden hulls with iron plates attached to the bottom to protect the hull when the ship encountered ice. Even in the middle of a bleak winter, cargos had to be shipped, and people had to travel from one town to the next. The easiest way to do this was by water. The *City of Duluth* was one of the ships that was involved in this trade.

Hull no. 125278 first tasted fresh water in 1874 with a maiden trip between Cleveland, Ohio, and Duluth, Minnesota. Initially owned by Solomon Gardner, the *City of Duluth* was one of the most elegant vessels at the time. The following year, she was sold to Alvin C. Burt of Detroit and put on the cross-lake run between Chicago and Grand Haven. Leased to the Wards Lake Superior Line in 1876, she went on the Detroit, Michigan, to Lake Superior route. Sold to John Pridgeon in 1877, she began to run from Chicago, Illinois, to Duluth, Minnesota. In 1880, John Pridgeon sold part of his ownership to Charles Spencer. Then, in 1881, Spencer and Lyman Hunt of Chicago bought the remaining ownership of the *City of Duluth* from John Pridgeon.

In 1883, while the *City of Duluth* docked in Bayfield, Wisconsin, she became a minor player in the tragic tale of the steamer *Manistee*. While at Bayfield, the *Manistee* came into port on Sunday, November 10, 1883. Lake Superior was in a nasty mood and had severely mauled the *Manistee* when she left Duluth the previous day. The storm turned into a full-blown gale and did not let up. The *Manistee* made several attempts to leave port and was beat back into the harbor each time. Finally, on Thursday, November 15, there appeared to be a break in the gale, and the *Manistee* decided to leave port that evening around 8:40 p.m. Before she left, many of her passengers decided to leave the *Manistee* and board the *City of Duluth* or the steamer *China*. Perhaps they had a feeling about the *Manistee* or just felt that these other boats had better accommodations and construction. Either way, when the *China* and the *City of Duluth* arrived at Portage the next day, they expected to see the *Manistee* there in port. Initially, it was thought that she might have returned to Duluth, but when the steamer *Hurd* arrived and reported that she was nowhere to be found, tugs were sent out to look for her. Finally, on November 25, pieces of the ship were found floating in the vicinity of the Apostle Islands, indicating that the *Manistee* had gone to the bottom. None of the crew or passengers was found. Fate has a strange way of playing its hand. Whether the passengers had intuition or just plain luck, many of them felt both shock and relief when they heard the news, feeling fortunate that they had chosen to leave the *Manistee* for passage on the other steamers.

In 1886, the *City of Duluth* was taken to Buffalo, New York, to be lengthened. She now had a length of 202 feet, with a beam of 36 feet, and could carry a cargo of 912 net tons, which gave her a depth of 13.5 feet. The lengthening and added depth helped seal her fate later. She was bought and sold over the next few years and finally bought by Albert Spencer of Waukegan, Illinois; Lyman Spencer of Buffalo, New York; Samuel Leopold of Chicago, Illinois; and Mr. Austrian of Hancock, Michigan, in April 1896. These gentlemen formed the Lake Michigan and Lake Superior Transportation Company. In 1897, Graham & Morton in St. Joseph, Michigan, chartered her to run between Chicago and St. Joseph that winter.

Early in the morning on January 26, 1898, J.H. Graham paced his office. Being the president of Graham & Morton was no easy task, and worrying about the passengers and his crew was one of the burdens of his career. His stress seemed to intensify after November 1 of each year, when the lakes start to show their Mr. Hyde personality in vicious cold storms and gales. During the summer months, the lakes display their Dr. Jekyll side in the form of warm, soft breezes and relatively smooth sailing. Remembering the loss of the *Chicora* several years earlier, Mr. Graham sent a telegram to Captain Donald McLean of the *City of Duluth*. He had hesitated too long with the *Chicora*, and the telegraph missed being delivered to Captain Stines by only ten minutes. The *Chicora* had already set sail and was last seen leaving the harbor in Milwaukee by the boy who had tried to deliver the telegram. After that, the *Chicora* was never seen again. Even today, her sinking remains a mystery.

The *Chicora* and the *City of Duluth* shared many similar circumstances, including a falling barometer that indicated a storm was brewing and a departure from a port on the west coast of Lake Michigan. However, there were also some differences. The *Chicora* was set to leave Milwaukee instead of Chicago. Also, the *Chicora* was owned by Graham & Morton, whereas the *City of Duluth* was leased. The *Chicora* was not insured, so when she was lost, it negatively affected the company financially. The telegram sent to Captain McLean of the *City of Duluth* instructed him not to leave Chicago until the afternoon of that day and to wait until 11:30 p.m. to consider sailing. That way, the *City of Duluth* would still have daylight to help her get into St. Joseph Harbor. Mr. Graham knew that the sandbars at the entrance, right outside the breakwaters, were constantly changing. That morning, the *City of Traverse* had ripped off her rudder trying to negotiate the entrance. He was greatly relieved to learn that Captain McLean received his message. Little did he know at the time that Mr. Austrian, one of the ship's owners, was going to overrule the message's instructions. Finding the *City of Duluth* still tied up to the Chicago docks late in the morning, Captain McLean was told that he was too cautious, and Mr. Austrian did not care about Mr. Graham's telegram.

Leaving the dock around 2:15 p.m. that afternoon, the *City of Duluth* had a reasonably easy run across the lake. But as the ship got closer to St. Joseph, the captain began to question the wisdom of trying to enter the harbor at night, especially after the experience they had about a month before. The *City of Duluth* had snagged a sandbar and was stuck fast in the waves and ice before the tug *Andy* came out and tried to pull her off. After the tug labored without any noticeable results, it waited until Mr. Graham sent out the *City of Louisville* to offload the *City of Duluth*'s cargo, which made her light enough for the tug to pull her off the sandbar.

Closer to the harbor of St. Joseph, the wind and waves began to pick up. Traversing the mouth of the river at St. Joseph can be a heart-pounding experience. The wind will try to blow a boat one way, while the currents try another way. Additionally, the waves generated out in the lake will meet the outgoing river current and create a jumbled assortment of waves, tossing any boat around.

The *City of Duluth* arrived off the entrance to St. Joseph Harbor around 8:15 p.m. As Captain McLean slowed the ship to align her to the entrance, the passengers could see the beckoning city lights and felt a sense of security as they approached the north and south piers. Fighting a moderately heavy wave action and a strong southwest wind, Captain McLean struggled to keep her on course. About the time Captain McLean started think that he might make it between the piers, the *City of Duluth* was jolted to a sudden stop as her bottom plunged into a newly formed sandbar. He hit the same sandbar that bit off the rudder of the *City of Traverse* earlier that morning. As the crew felt the sudden stop, Captain McLean ordered the *City of Duluth*'s engines to full speed, hoping that he could create enough energy to carry the ship over the sandbar on the lift of the next wave. Black smoke poured out of her smokestack, and instead of lifting the ship with the thrust of the thrashing propeller, the bow of the ship drove deeper into the bar.

Captain McLean ordered the engines shut down and waited for help from shore; he knew that the ship was firmly stuck. For about an hour, the steamer sat there, and finally, around 11:00 p.m., she broke free of the sandbar and became a plaything for the wind and waves. Spinning her around 180 degrees with her bow pointed back out into the lake, she grounded again about three hundred feet off the north pier. As the boat settled in the shallow water, the shattering of glass and the groans of the dying ship were audible. The *City of Duluth* started to crack in two about her midsection. With the report of loud cannons, her starboard's supporting arch snapped and fell into the lake. As the passengers had first started to panic, Captain McLean and his crew tried to calm their fears, reassuring them they were not in danger. As the water entered her hull, the crew extinguished the fires in the boilers, leaving the ship without heat. Those on board included the following:

CITY OF DULUTH'S PASSENGERS

Name	*City*
August Kernwein	St. Joseph
Mrs. Frances Sowers	Benton Harbor
Mrs. Wm. Tryon	Royalton
Mrs. Wm. Clark	Watervliet
H.J. Ray	Watervliet
Robert Tripp	South Haven
Samuel Wellman	Detroit
W.O. Slight	Baroda

H.L. Lemon	Baroda
Peter Ahlers	Chicago
Harry Sowers	Benton Harbor
Peter Fisher	Chicago
R.D. McCuskey	Sister Lakes
Walter Keigley	Eau Claire
Thomas Hagerman	Benton Harbor
E. Pott	South Haven

City of Duluth's Crew

Name	*Rank*
Donald McLain	Captain
Bert Simons	First Mate
Jas Quinlan	Second Mate
Chas Andrews	Wheelsman
M.J. Morrison	Wheelsman
O.A. Shauman	Clerk
Edward Nolan	Steward
Henry Chalk	Chief Engineer
Ernest Narjal	Second Engineer
Wm. Stetwelt	Fireman
Ed Barnes	Fireman
Dick Peck	Oiler
J.L. Hanson	Watchman
Sam L. Johnson	Watchman
John Kline	Coal Passer
Joseph Mitchell	Cook
Roy Sariline	Cabin Boy
Thomas McGrain	Porter
Peter Lotwine	Deckhand
Alex McKee	Deckhand
Chas Brown	Deckhand
Chas Fox	Deckhand
Morris Feeley	Deckhand
Ed Smith	Deckhand

With forty souls on board, Captain McLean was not confident about what he was telling his passengers about their safety. He knew that the *City of Duluth* was in more trouble than what he had indicated.

When Mr. Graham was telegraphed about the departure of the *City of Duluth* from Chicago, he went down near the south pier to watch for her. When he saw the ship hit the bar, he asked the captain of the tug *Andy* to get up steam. An hour later, when the ship was heading to her final resting place, the tug *Andy* headed out into the waves to try to pull the passengers and crew off the wreck. At the same time, he sent a runner to Keeper William Stevens of the U.S. Life-Saving Service to get his crew ready in case the tug failed in the wild waves around the wreck.

Captain William L. Stevens became the keeper of the lifesaving station in St. Joseph, Michigan, in 1879 after Keeper Joseph Napier resigned from the service. Born in St. Joseph, Michigan, on December 5, 1851, Stevens worked in the family business and captained several ships. Stevens started his career in the service under the tutelage of Joseph Napier. In 1879, at the young age of twenty-eight, he was named keeper of the station in St. Joseph. During the short time Joseph Napier was keeper, he displayed admirable dynamism. William Stevens had enormous shoes to fill. He remained in charge of the lifesaving station until retiring in 1914. In his thirty-five-year career as keeper, Captain Stevens and his crew rescued an estimated three hundred or more people from shipwrecks, overturned rowboats and other calamities on Lake Michigan.

Not only did William Stevens train a great crew of lifesavers, but one of the unofficial members of his crew was his wife, Ella. In the years he was in the service, they both lived at the station year-round. Typically, most stations around the Great Lakes close at the end of the shipping season, and the surfmen work other jobs during those months. The keeper usually goes to a permanent home, but not William and Ella—they stayed at the station all year. They made hundreds of friends, many of them from Chicago. Like they still do now, Chicago residents come across Lake Michigan to enjoy the cities of St. Joseph, Benton Harbor and the east coast. Naturally, many of these people would eventually make good friends with William and Ella. Ella would always have a pot of coffee, iced tea and food ready. No one left that lifesaving station hungry. Not having any children of her own, Ella became the unofficial mother to many of the younger men serving under her husband.

Many times, Ella accompanied the crew on a rescue. She would build a fire on a beach or in the station to warm the people they rescued from the bone-chilling waters of Lake Michigan. In addition to a pot of coffee, Ella had dry clothes ready for the men to restore the heat they lost when dealing with the lake. During their years of service, Captain Stevens and

Ella produced more captains for other lifesaving stations along the east coast of Lake Michigan than any other keeper. Captain Morgan of Grand Haven, Captain Mathews of South Haven, Captain Lysaght, Captain Morrison of South Chicago, Captain Johnson of South Haven and Captain Flynn of Point Au Sable were all members, at one time or another, of the St. Joseph Life-Saving Station. There, they learned the essentials in the art of being a storm warrior.

When Mr. Graham sent word of the *City of Duluth* to Captain Stevens, the station was already closed for the season. Additionally, the crew had been paid for the year. Nonetheless, Captain Stevens was able to pull many of his dedicated men back to the station. He also recruited help from local citizens and the crew of the *City of Traverse*, so they now had the men necessary to perform the rescue. It took several hours to get the men to the station, and in the meantime, Keeper Stevens formulated a plan based on the wreck's location. Ice floes were already floating in the water, and Keeper Stevens decided that going out in the lifeboat was out of the question because the waves would turn those ice floes into battering rams that could sink or damage the lifeboat. The next logical plan was to get the equipment necessary to fire a line out to the wreck and bring the survivors to shore with the breeches' buoy. An initial line had to be shot out to the wreck to get the breeches' buoy out to the ship. With that line, the ropes and pulleys needed for the breeches' buoy would be pulled out and secured to the wreck. The other end would be secured someplace on shore. An initial line was shot out to the shipwreck using a Lyle gun. Even though it was small, it still weighs about 202 pounds.

Although two people could carry the Lyle gun, it was still not considered light. Additionally, the projectiles, rope, gunpowder and other necessary equipment would have to be brought to the rescue site. For this particular rescue, the men would have to carry all this equipment. Keeper Stevens had looked over the different locations and thought that the best one was at the end of the north pier. This location was the closest to the *City of Duluth* and would be well within the range of the Lyle gun. Keeper Stevens's concern was that he needed about twenty-five people to carry the gear the half mile out to the end of the pier. There was considerable risk with that many people involved in such an endeavor. Waves crashing on the north pier's surface covered the surface in snow and ice, and a twelve-foot hill of snow and ice blocked them from the end of the pier. For anyone venturing out on the pier, there would be a likely chance an individual would slip and fall into the lake. Then his life would be measured in minutes because the icy water would quickly induce hypothermia. The victim would have to be pulled out of the freezing water immediately and placed into an area to restore his body temperature and dry his clothes. Therefore, the move to the end of the pier would have to be done with extreme caution.

As Keeper Stevens watched intently, the tug *Andy* steamed past to make the first attempt to rescue the people of the *City of Duluth*. He had his men to start hauling the equipment out to the end of the pier, selecting each step carefully and warning them constantly to watch their step. He watched the tug near the end of the breakwater and saw her pitch bow up in the turbulent waves at the mouth where the river meets the lake. The combination of the wind and waves was extremely hazardous. Also, the waves on the lake came from one direction and battled against the strong current coming out of the St. Joseph River. This produced waves in multiple random directions, regardless of the wind. Once past these waves, the tug *Andy* would only have to contend with the sharp waves on the lake. The tug moved in a big arc and attempted to come alongside the *City of Duluth* and get a line to her. However, the wind blew the rope away from the wreck. The *Andy* attempted three times to secure a line to the *City of Duluth*, and three times the weather conditions proved too much for the tug skipper. Finally giving up, the *Andy* returned to the safe and calm waters of the harbor, leaving the crew and passengers on the *City of Duluth* to their fate.

Now it was Keeper Stevens's turn to try to rescue the people off of the wreck. The first obstacle was to get the 202-pound cannon and all the ropes, pulleys and associated gear over a twelve-foot wall of ice and snow at the end of the pier without having anyone fall in the icy waters of Lake Michigan. Ella Stevens was among those in the group. She was not going to allow her boys to risk their lives without her being a part of it too. First, some of the men reached the top of the ice wall, and then with that extra muscle, they safely got the Lyle cannon over it. Then they hauled the equipment first, followed by the lifesavers and then the people who came along to help.

The last man over the wall was Keeper Stevens. When they finally completed their half-mile journey to the end of the pier through snow and ice, they saw that the *City of Duluth* was in a desperate situation. She had sunk low enough that the water reached her main deck. They hurriedly assembled the equipment on the windswept pier. The wind was coming straight off the lake, making the icy spray feel like sharp glass stabbing their exposed faces. It did not slow them, as hours of intense drills had made this exercise second nature. The group prepared the cannon to fire.

In 1898, the north pier in St. Joseph was dramatically different than what you see today. Now, there is only a small lighthouse at the end of the pier,

Wrecked *City of Duluth*. *Alpena County George N. Fletcher Public Library, Great Lakes Maritime Collection.*

St. Joseph's north breakwater. *Donated by Richard DeClercq.*

but in 1898, there was a larger lighthouse and a fog signal building. There is a lot more area today on the end of the pier than in 1898. One of the first tasks for Ella Stevens was to go into the fog signal building and start a fire to warm up both the survivors and lifesavers. This fire also allowed Ella to start some coffee and soup.

Finally, the cannon was ready to fire its first projectile. It was loaded into the muzzle and the leader line attached. The Lyle gun emitted a resounding *bang*, and the first shot sailed into the night. But the wind blew it out of reach of the people on the wreck. Making slight adjustments, Keeper Stevens lined up the gun again. The second shot rang out, and the projectile and line sailed perfectly across the *City of Duluth* within easy reach of the survivors. The crew quickly hauled in the breeches' buoy lines and attached them to the stern rudder post. Once secured, they pulled the breeches' buoy out to the wreck. It was now around 10:00 p.m., and the rescue of the survivors could begin.

Once the breeches' buoy reached the wreck, Captain McLean intended to send one crew member, preferably his first mate, to test the ropes and ensure that everything was safe. But August Kernwein, of St. Joseph, was so anxious to reach dry land that he kept pushing himself in front of everyone, including the ladies. So, the captain decided to get rid of this problem passenger. He asked him to go first to make sure everything would work. As August got into the buoy and started toward the north pier, the ropes could not be tightened enough. Thus, the passenger was waist-deep in the freezing cold water for about half the trip. Once he reached the pier, through chattering teeth, August informed Keeper Stevens that he would have to send one of his lifesavers out to the wreck. An elderly lady, Mrs. Tryon of St. Joseph, was disabled and could not walk. Captain McLean was extremely concerned about sending her in the buoy alone and requested a lifesaver to ride along with her. Chas Roberts volunteered and got into the buoy for the ride to the wreck.

Once there, Chas Roberts informed the woman that Captain McLean was giving her his personal attention. When the ship first wrecked and everyone was heading to the upper decks, Captain McLean learned that she was unable to climb the steps to the upper deck. He picked her up, carried her and made sure that she was as comfortable as possible. When the breeches' buoy was ready, Captain McLean again picked her up gently and put her into the buoy. Chas Roberts also climbed aboard and made sure she would not fall out. Upon reaching the north pier drenched by the icy water, Chas Roberts took her in his arms and carried her to the fog signal building, where Ella Stevens had dry clothes, a warm fire and hot soup to revive her. As he was taking care of Mrs. Tryon, the buoy was sent out again and again without a problem until Robert Tripp of South Haven was being pulled in. A chunk of ice took away one of the lines, and for a moment, it seemed that they had lost Robert Tripp. With great effort, they untangled the lines and got Robert Tripp safely on the north pier. The remaining crew and passengers kept coming ashore until 4:00 a.m. The last person off the ship was Captain McLean. He was shuttled immediately into the fog signal building to recover from the dunking he had in the cold waters of Lake Michigan. The captain could barely make out the outline of the wreck in the gloomy darkness and crashing waves. He was undoubtedly thinking, "Mr. Austrian, I told you we should have heeded Mr. Graham's telegram."

Henry Chalk, the chief engineer of the *City of Duluth*, was understandably upset about the loss of his ship. He was sure that she was beyond returning to service. Mr. Chalk had been with her since the beginning of navigation in 1880. Unlike the pilothouse officers, engineers tend to become attached to their ships and rarely leave them. When they do lose a ship, they grieve as if they have lost a loved one. Still, he had to console himself that not one life was lost because of her wrecking.

While everyone was recovering, Mrs. Tryon was taken to the local hospital. She was having difficulty recovering from her ice water dunking and went back to Chicago for treatment. The passengers and crew who did not live in St. Joseph or the neighboring town of Benton Harbor were put up in the hotel Higby by Mr. Graham. Even though the local reporters detailed the names of the gallant crew, they downplayed Ella Stevens's heroism. Following is a list of the crew:

U.S. LIFESAVER SERVICE, ST. JOSEPH

Captain Stevens	Richard Stines
Albert Carter	Frank Fowler
Chas Roberts	Fred Alden
William Moyer	Chas Burkhand

People Who Assisted in the Rescue

Captain Lloyd	Captain D.R. Platt
Captain Mollhagen	John Mollhagen
William Waterson	Amil Risto
Fred Griffin	Harry Hughson
Dr. Scott*	Ella Stevens

*Dr. Scott was on hand to help treat those who needed medical assistance.

There were also many others whose names the reporter did not record.

As many of the survivors and rescuers were drying their wet clothing in the fog signal building, a reporter singled out one of the elderly ladies sitting by the fire in the boiler room, Mrs. Marie Clark, age seventy-five, Mrs. Tryon's mother. She had gone to Chicago to care for her invalid daughter. As the reporter assisted Mrs. Clark in taking off her wet shoes and putting them close to the boiler to dry, he noticed that she was in a very cheerful and philosophical mood. He asked her about her experience and if she would ever take another trip. She replied:

> *At no time was I badly frightened and my experience of last night would not keep me from making a lake voyage today, were it necessary. I knew that all that could possibly be done was being done. The only thing that caused me any fear or tended to make me nervous was the breaking of the glass and the splintering of the wood and the flying about of these broken pieces of the ship and cabin. The place was rendered very noisy, and uncomfortable, but I am glad to say that I have experienced no injury and am only slightly fatigued.*

The next day, after the excitement and seas had calmed down, Mr. Graham and others from the Graham & Morton offices went to examine whether the *City of Duluth* could be salvaged. They found her badly broken up below decks, and her engine had broken loose of its mounting. She had also broken in the middle, and there was a large hole visible below decks. They were convinced that a lot of her cargo could be salvaged but noted that some of the bags of flour were in the water and probably would be worthless. When he returned to his office, Mr. Graham telegraphed the owners of the *City of Duluth* and the insurance underwriters of her cargo that Graham & Morton had officially abandoned their interest in her and her cargo. He also ordered the wrecking tug *Morford*, which was sent over from Chicago the previous night, to return to Chicago. Now it was up to the insurance companies to decide how to recover anything from the wreck.

The insurance underwriters also determined that the *City of Duluth* was a total wreck and started preparing to salvage what they could. Her owners, Lake Michigan and Lake Superior Transportation Company, had insured her for $70,000, and Graham & Morton, which owned the cargo, had insured it for $10,000. So the insurance company wanted to salvage as much as it could from the wreck to reduce its losses. It hired a tug and barge, offloaded much of the cargo and stored it in a Graham & Morton warehouse. The insurance company faced another problem: the shipwreck had become a popular tourist attraction. St. Joseph was booming as people clamored to see the shipwreck. The ice that had formed out to the wreck enabled people to board her easily. Everyone wanted a souvenir and picked up anything that they could effortlessly haul away. All of the *City of Duluth*'s silverware and china vanished. Luckily, the lifesavers had already taken the crews' and passengers' personal property immediately after the rescue. Next, some of the cargo that had appeared to be damaged started to disappear. People in St. Joseph and Benton Harbor found that the wet bags of flour had formed a shell, and their centers still held good flour. People in the area did a lot of baking that winter! Once the insurance companies were done with their salvage, the U.S. Army Corps of Engineers surveyed the wreck and declared it a hazard to navigation. It took bids for removing the wreck from its current location. A furious storm swept across the lake on January 31 and destroyed the wreck even further. The waves and battering rams of ice washed the cabins amidships overboard.

In May 1899, the wrecking schooner *Judd* was tied up next to the wreck. Captain Jix of the schooner reported that he removed the engine and boiler from the wreck. This involved tearing off the top deck to more easily access the engine room. Pieces of the deck and hull were hauled out into the lake and sunk. The engine, boiler and most of the starboard side, with its arch, were hauled two miles down the lake and sunk. In August of that year, the remaining hull pieces were taken out farther into the lake and sunk. The salvagers dumped the last remains of the wreck about two thousand yards straight out of the mouth at St. Joseph. This location is marked on navigational maps. So ended a proud ship and an extraordinary rescue. The capabilities and training of the crew in St. Joseph are a credit to Captain Stevens as the keeper of the lifesaving station for that success. Some of this praise should go to his wife, Ella, for all her hard work keeping those boys warm and fed during this and other heroic rescues.

Twice Rescued

Often, the adventures on Lake Michigan end tragically. Such was the case of the schooner *Arab* and her crew. If the schooner's owner had abandoned the aged ship, instead of insisting on repairing her, two other ships, along with the *Arab*, would not have been lost. Additionally, several men's lives would have been spared. How often have we wished, with the advantage of hindsight, that we had made a different decision than the one we did? As he was trapped underneath a heavy steam pump in a schooner plunging into the bottom of Lake Michigan, engineer William Kelly probably regretted his decision.

On October 29, 1883, the small seventy-eight-foot schooner *Arab* was at Starke's Pier near Frankfort, Michigan. She was topping off her cargo of lumber, piles, slabs and cordwood in preparation to sail. Built in 1854, the twenty-nine-year-old schooner had seen better days. Only able to haul a mere 209 tons of cargo, she did not yield a lot of profit to properly maintain the aged ship. The owners only opened their pocketbooks when the schooner needed major repairs to keep her off the bottom of the lake. Her equipment was worn, she leaked constantly and she required constant vigilance and pumping to keep the water inside at a manageable level. Captain Starke had a tough time retaining an entire crew of four for the schooner during the shipping season. Little did he know that this would be his last run in the *Arab*. When all the preparations were complete, he got underway for his home port, the city of Milwaukee. Once out into the open lake and setting a southwest course, Captain Starke noticed that the lake was becoming boisterous, and the farther he went on his course, the rougher the waves became. Just northwest of the St. Joseph Harbor, the ship acted sluggish to the wheel, so he ordered his four-person crew to toss some of the deck load overboard. Then, without warning, he heard something snap, and a quick investigation revealed that the centerboard, which kept the *Arab* stable, had snapped off. The closest port for safety and repairs was the St. Joseph Harbor. Changing to a southeast course, Captain Starke headed his schooner in that direction, knowing that there was an experienced lifesaving crew manning the station if they got into trouble. At about 7:30 p.m., on Halloween night, as the *Arab* was nearing the safety of the harbor and riding low in the water, she slammed into a sandbar. The lookout in the station's tower saw that she was heading for trouble and, in an effort to warn her, lit a hand torch. He either lit the torch too late or Captain Starke was too busy to see the warning. The *Arab* bumped over the sandbar, missed the entrance completely and settled by the foot of the south pier. It sat near what is now known as Silver Beach. Captain Stevens and his trained crew responded quickly and managed to tie the wounded *Arab* to the south pier. He rescued Captain Starke and his four men by getting them onto the pier. Captain Stevens and his crew took all five men to the station to dry their clothes and decide on a plan. They had to contact the *Arab*'s owners and find out what they wanted to do with the ship.

Captain Stevens, looking at her size and age, was sure that the owners of the *Arab* would abandon her where she sat at the end of the pier. He was astonished the next day when Captain Starke requested that his men help patch and pump her out for towing across the lake. The owners had already chartered the tug *Protection*, and she was on her way to the harbor. Captain Stevens and his men worked with the *Arab*'s crew and the crew from the *Protection*. They installed one steam pump but found that the water from the many leaks overwhelmed it. Even though there was only the one pump, the crew thought they had patched enough of the *Arab*'s leaks so she would not have trouble floating. They got her about fifty feet down the south pier in St. Joseph when the water rushing in overwhelmed the pump—again she settled to the bottom. Captain Stevens was more than upset that his men were wasting their time on such a derelict ship, especially since Captain Starke could have removed the cargo sitting atop most of the leaks. The wrecking master who came with the *Protection* decided to install a second, bigger pump on board the *Arab*. Again the men began patching what leaks they could find and hoped that the pumps could handle the water coming in from other leaks under her cargo. By the time they had fixed enough of

Tug protection. *Alpena County George N. Fletcher Public Library, Great Lakes Maritime Collection.*

the leaks with temporary patches, more than a week had passed, and the tug started pulling her toward Racine, Wisconsin, about 7:00 p.m. on November 10. As Captain Stevens saw the *Protection* leaving with the *Arab* at the end of a five-hundred-foot towline, he hoped never to see the *Arab* or her crew again. At least, with the moon softly lighting up the easy swells of the lake, they would have a calm night to get most of the way across the lake.

The tug *Protection* had eight men on board: captain, pilot, engineer, two firemen, deck hand, cook and the wrecking master. The schooner *Arab* had ten men on board: Captain Starke, first mate, steward, three seamen, two engineers and two firemen to handle the steam pumps. Shortly after they got out into the calm lake, William Kelly, one of the engineers, decided that they didn't need both pumps, so he shut one down. This unwise decision was going to cost him.

All night, as the *Protection* pulled the *Arab* on a northwest course, everyone on the tug was relaxed, lulled by the bright moonlight, slight breeze and softly lapping waves. So far, everything was going well. Many people who make a living on or near the lakes will tell you that you need to be vigilant about the weather and everything on board a ship. There's no telling when something that seems insignificant will cause you significant problems.

A frantic sailor insisted to engineer Kelly that there was a lot of water forward. He rushed to get the second steam pump working, knowing that once a fire starts in the boiler, the steam pressure has to rise high enough that the pump will begin to expel water. The water continued to rise, and with the momentum of the *Protection* and the steam pumps' extra weight on the bow of the *Arab*, the old schooner rolled over on her side. As she rolled, the steam pump that engineer Kelly was working on toppled over on him, pinning him to the deck. All the other crew members raced for their lives to the aft of the schooner. The men on the *Protection* heard the men's loud screams and saw the old schooner start to sink. Frantically, the pilot on the tug ordered all stop and then reverse without thinking about the long towline. As the towline wrapped tightly around the *Protection*'s propeller, her engine came to a sudden stop. As the tug's crew tried to figure out what had happened, the captain heard the crew of the *Arab*, who were clinging to the aft end of the schooner, screaming and pleading for help. The captain of the *Protection* ordered their small lifeboat launched to rescue the sailors desperately clinging onto the stern of the *Arab*. When the crew of the *Arab* came on board the now-disabled tug, they told the story of the unfortunate engineer who was on his way to the bottom of Lake Michigan. With nine men from the *Arab*, plus the eight men already on board the tug, the priority became getting the towline untangled from the propeller—that way they could do something other than drift to the whim of wind and waves.

Lake Michigan, like all the other lakes, has a distinct personality. Most of the time, in the summer months, especially nowadays, it is a playground for those who swim, boat, fish and enjoy its pleasures around the shore. But it also can have an evil side, especially for those who make themselves vulnerable to its currents and waves. These days, there are about twenty-eight deaths annually from drowning around Lake Michigan. These deaths are of people who do not understand the lake's unpredictability and treat it like a big pond. They jump off piers that jut out into the lake without knowing that there are giant boulders just under the surface. Or they get caught in rip currents and, with an adrenaline rush of panic, fight against them until they have no energy left and disappear under the surface. These folks become numbers in the long list of statistics. As the tug *Protection* wallowed in the gentle waves, her crew desperately attempted to free her propeller. However, at that moment, Lake Michigan knew that it had snatched another victim.

It started with a slow increase in the wind velocity, barely noticeable to those on board. The crew was concentrating on the towline and the propeller. But the waves grew bigger as the vessel helplessly drifted. By 10:00 a.m., the sinister change in the wind and waves had forced the work on the propeller to stop. The wind had switched direction to the northwest and started to blow hard. The tug rode amazingly well in the waves without any power. About an hour later, she spotted the big steamer the *H.C. Akeley* of Grand Haven. With the frantic pull of the whistle, she signaled to the *Akeley* that she was in trouble.

The *Akeley* had just put out of Chicago several hours before, heading for Buffalo, New York, with a cargo of corn. Hearing the distress calls from the foundering tug, the big steamer changed her course to render aid. Pulling up next to the tug and hearing from her crew the nature of their problem, they secured a towline and proceeded to pull the tug toward Grand Haven. The *Akeley*, with a crew of eighteen, was indeed a welcome sight. But the farther they went, the more the wind and waves increased. By that afternoon, Lake Michigan was blowing a full gale. The signal office in Grand Haven recorded the wind velocity at fifty-two miles per hour. Even though the tug rode the waves reasonably well, the big steamer started showing signs of stress with the tremendous waves and pulling the *Protection* behind her. As she rolled heavily, the men on the tug could tell that the *Akeley*'s cargo had shifted and was listing. Around 7:00 p.m., the strain of pulling the tug through mountainous waves washing over the shifting cargo caused the steering gear that controlled the *Akeley*'s rudder to become disabled. Soon after her rudder stopped working, a feed line to her port boiler burst, leaving her without an engine. The situation on the *Akeley* was going from bad to worse. Powerless, the *Akeley* started to drift helplessly in the waves along with the *Protection*. The *Akeley* still carried sails on two masts, and her captain ordered them raised to bring her out of the trough between the waves. The sails lasted only a short time before the wind blew them to rags. They drifted this way into the night, and around 4:00 a.m., one of her lifeboats was swept over the side. Then, at 10:00 a.m., the big steamer lost her smokestack. Without the smokestack creating an airflow through her firebox in the boilers, there was no hope of getting her engines running again. All this time, the tug *Protection* was still tied to her.

Of the two ships, the *Protection* was handling the tempest better than the *Akeley*. The tug was riding the waves with little difficulty. Workers continued trying to free the *Protection*'s tangled propeller. Around 6:00 p.m., the men on the tug tested whether they had propulsion and found that the waves had sufficiently alleviated the problem. When the waves started to moderate, the tug informed the men on the *Akeley* that they would head to South Haven. They figured by dead reckoning that they must be about thirty miles away from there. The plan was to get more coal and then return to tow the *Akeley* into Grand Haven.

H.C. Akeley floundering. *Alpena County George N. Fletcher Public Library, Great Lakes Maritime Collection.*

Shortly after leaving, the men on the tug found that the towline was still a problem, so they shut down the engine and drifted with the wind and the waves. She drifted this way, and at about 9:00 a.m. on November 13, they were north of Saugatuck and about a half a mile out into the lake. Dropping anchor and seeing many people gathered on the beach, they took a piece of awning and tied it to an oar. They used it to signal the people on the beach that they were in distress, hoping a tug out of Saugatuck would tow them into safety. But no one was willing to risk taking on the tremendous seas and perhaps become a casualty too. One of the difficulties was that the entrance to Saugatuck is only forty feet wide, and it would be nearly impossible to tow a boat into the harbor under such forbidding circumstances. With each wave tossing the *Protection* around, people on the beach could see the windblown individuals on deck hanging on for dear life. After a hasty conference, some of the leading citizens from Saugatuck decided that a rescue of the *Protection* and her crew was not within their capabilities, and they needed professional help.

They telegraphed the lifesaving station in Grand Haven, which was only thirty-three miles north of Saugatuck. Shortly after sending the telegraph, they got a reply, informing them the lifesaving crew was busy with another wreck and would not be available for some time. The next choice was the St. Joseph Life-Saving Station about sixty miles south of Saugatuck. Immediately, a telegram was dispatched to Captain Stevens. Captain Stevens tried to send a telegraph to the district supervisor at Grand Haven for permission to travel that distance for this rescue. The captain received a quick reply saying that the Grand Haven crew was involved in another rescue. He decided to respond to the Saugatuck request with utmost dispatch, even though it was a great distance for his crew to travel. Keep in mind, these distances were much more difficult to traverse then. The modern roads and highways that we are so accustomed to now did not yet exist. The two fastest forms of transportation were by steamship or railroad. Since there was a gale with monstrous waves blowing, a steamship was out of the question. So Captain Stevens's only option was the railroad. After several hours of intense activity, Captain Stevens arranged with the local Chicago and West Michigan Railroad agent to transport his lifesaving equipment to a destination near Saugatuck. He then telegraphed the citizens in Saugatuck that he and his crew were on their way. Leaving only one man, his number one surfman, in charge of the station, Captain Stevens loaded the equipment necessary for the rescue.

This was a daunting and dangerous task, considering it had to be taken across the river from the lifesaving station to the railroad terminal on the other side. Waves coming off the lake were being funneled into the river, which made it hazardous to cross by boat. They first strung a line across the river, thus adding a measure of safety for ferrying equipment across the river. With the Lyle gun, its cart and all other necessary equipment to load, the crew had to make several trips in the boat to transport all the equipment to the train station. This work left only ten minutes for loading the gear onto the baggage car. Usually, it would be placed onto a flatcar, but since none was available, they needed to take the wheels off the cart to get the equipment through the baggage car's door. All hands worked, plus a few of the railroad workers. They were soon all aboard, and by 12:55 p.m., the train was racing out of St. Joseph, barreling down the tracks headed for Saugatuck.

By 3:00 p.m., the train had reached Richmond on the banks of the Kalamazoo River, the closest point to Saugatuck. The tug *Gauges* had been sent upstream from Saugatuck to meet Captain Stevens and his crew. In a few minutes, all the lifesaving equipment was deposited onto the tug. With the weight of eight lifesavers, the tug's crew and the equipment, the tug rubbed bottom several times on its thirteen-mile downstream trip. By 5:00 p.m., they had landed the equipment and men near the lighthouse and moved the gear abreast of the wreck on the beach. Still riding at anchor, Captain Stevens's opinion was that the *Protection* was out of range of the Lyle gun. Since the tug and her crew appeared to be in no immediate danger, they decided to get the Lyle gun ready but also wait to see if the anchor on the tug would start to drag. While waiting for the lifesavers to arrive, locals twice took a Mackinaw boat to try to rescue the men. Both of these attempts failed due to the exceedingly high waves that lobbed the boat back onto the beach. At about 6:00 p.m., the wind made a horrendous shift to the northwest and abundant snow fell.

Around 9:00 p.m., the *Protection*'s whistle started to wail, and Captain Stevens knew that the tug was dragging her anchor. The men calculated that her drift was roughly south to southeast, and she would eventually hit near the beach south of the entrance to Saugatuck. Seeing that help was not forthcoming, the men on the tug relinquished any hope of being saved. The tug members shook hands and resigned themselves to a cold and torturous death. However, the lifesavers had not given up yet. They repacked their gear and, with the excited crowd, moved the equipment back to the river. The government tug was waiting for them there, and it ferried the equipment and men across the river to the beach. Once across, a lot of driftwood and snow hindered their progress. A large pond presented another obstacle. With ropes tied to the cart and the backs of the excitable locals, it took more than an hour to get to where they thought they could set up their equipment, which was the place *Protection* would hit the outer sandbars. A short time later, the men on the tug felt the first jolts of her hitting the first sandbar. Immediately, the waves started to sweep over her and tore away part of her pilothouse. About the same time, William Grace, one of the young firemen on the tug and a favorite of many of the crew, was forcibly hit by a wave near the stern and washed overboard to his death. He had been warned by several crew members not to stand in such a perilous position, but he had ignored them. After several more jolts, the tug finally came to rest about two hundred yards offshore. The men felt a glimmer of hope when they saw bonfires on the beach. The bonfires' light illuminated the lifesavers setting up their equipment. As they were assembling the Lyle gun, a figure of a young man rose out of the surf, walked calmly over to one of the fires and started to warm himself. This man had donned a life vest on the tug and desperately plunged into the wild waves. Beating the odds, he swam through the freezing cold waves and survived.

When the tug's bow was pointed toward shore, the lifesavers saw their chance and loaded the Lyle gun. With a loud *bang*, the Lyle gun's projectile arched into the night, and the line dropped neatly on the stern of the *Protection*. But before the men on the tug could retrieve the line, the waves washed it back into the lake. The men brought the Lyle gun back to shore, loaded it again and re-aimed with a new projectile. With another resounding *bang*, the projectile arched into the night with the line trailing behind. As the line was pulled out into the lake, its main part draped neatly across the tug's midsection. The *Protection*'s men grabbed the line, and within a half hour, they had rigged the breeches' buoy; the first man was soon ready to be transported to shore. When he came ashore, the *Protection*'s movement in the waves slackened the lines that transported the breeches' buoy. About halfway to shore, the man was submerged up to his waist in the numbingly cold water. Being under water sapped his energy and caused the initial signs of hypothermia. Once on shore, he was bundled in blankets and rushed to the nearest fire to dry his clothes and restore his warmth. Given alcoholic stimulants, the man recovered from his dunking in Lake Michigan. All the bucking and moving of the tug made transporting each survivor difficult, binding the line, even with thirty or forty men hauling on them. Captain Stevens and another surfman waded out into the bone-chilling lake up to their waists to keep the lines from tangling. When Captain Starke of the schooner *Arab* came ashore, Captain Stevens was a bit taken aback that this was the second time within weeks that he was rescuing Captain Starke and his crew. Once all the men left alive on the *Protection* were safely on shore, Captain Stevens had time to contemplate the cost in men's lives because the tightfisted owners had tried to salvage a derelict like the *Arab*. He had no idea about the drama that was playing out in the middle of the lake on the *Akeley*, nor that the human cost would increase.

After the *Protection* left the *Akeley*, the *Akeley*'s crew engaged in frantic repair efforts. Without a stack or any means of propulsion, Captain Stretch ordered the anchor dropped, hoping it would bring the *Akeley*'s bow into the wind. As the weather conditions and those on board the *Akeley* further deteriorated, Captain Stretch ordered the last remaining lifeboat cleared away and ready to be launched at a moment's notice. At about 12:30 p.m. on Tuesday, the schooner *Driver* appeared on the horizon. The *Driver*'s captain, David Miller, saw the *Akeley* and ordered his brother Daniel to climb the rigging to get a better view. Scampering up the rigging, he shouted down from his perch that he could see her listing severely and noted that she probably would not stay afloat for long. He told his brother that he thought the ship was the big steamer *H.C. Akeley* and that they should get closer to render help.

As they moved closer, Daniel shouted to his brother in terror that the big steamer was foundering and her bow raised. Her stern started to plunge toward the bottom of Lake Michigan. Going as quickly as the schooner could go, they spotted the *Akeley*'s lifeboat. The *Driver*'s crew tried to adjust the sails to get closer to the lifeboat, but the winds and waves were not in their favor. Seeing that the survivors were having trouble nearing the *Driver*, Captain Miller asked for volunteers from his crew to man their lifeboat and help the men from *Akeley* make it to the *Driver*. One of the men stepped forward: Patrick H. Daly, known to the crew as "Paddy." He and Daniel Miller prepared the lifeboat. Lowering it, they maneuvered through the wave's stern first. This allowed them to get close enough to the *Akeley*'s lifeboat to transfer three survivors into the *Driver*'s lifeboat. Lightening the *Akeley*'s lifeboat allowed it to ride higher in the waves, making it easier to row toward the *Driver*. They threw a line to the *Akeley*'s lifeboat, which also helped them pull up on the *Driver*'s lee side. Once there, the men of the *Akeley* were pulled onto the *Driver*. But just about the time Daniel and Paddy were helping the last survivor on board, the waves capsized the *Driver*'s lifeboat, throwing both of them into the lake. Daniel and Paddy were able to swim to the tackle falls of the *Driver* and pulled themselves safely on board. When the men were finally on the schooner, they discovered that five crewmen from the *Akeley* were missing. Captain Starke, the first mate and three seamen decided to take their chances by staying on board the *Akeley*. Captain Starke made difficult life-and-death decisions. Unfortunately, Lake Michigan exacted payment in the form of human lives. Having lost five men on the *Akeley*, Captain Miller ordered his men to resume their original course to Chicago as his brother and Paddy went below for dry clothes and warmth.

The final result was that a simple salvage of a twenty-nine-year-old schooner in poor shape cost seven men their lives. Captain Stevens must have wondered if the owners of the *Arab* felt any guilt that their attempt to save a buck led to the deaths of these men. The U.S. Life-Saving Service awarded Daniel Miller and Patrick Daly gold medals and Captain David Miller a silver medal for their heroic rescue of the *Akeley*'s men. It definitely was a brutal rescue.

Daniel Miller's gold medal. *Tri-Cities Historical Museum-Grand Haven Michigan.*

The tug *Protection* sat on the sandbar until spring, when Captain R.C. Brittain of Saugatuck bought the salvage rights and pulled her off. The effort required the river steamer *Alice Purdy*, the passenger steamer *A.B. Taylor* and the fishing tug *Clara Elliott* to free her from the sandbar. She was finally towed into Saugatuck to be repaired, and in 1897, sixteen years after she towed the *Arab* off the beach in St. Joseph, she was abandoned in Saugatuck.

Sailors in Peril

The title of this tale is quoted directly from the October 27, 1898 edition of the *Detroit Free Press* because it aptly describes this story.

Feeling his age in his bones, the veteran sailor made his way to the breakwater in Michigan City, Indiana, where a large group of townsfolk had gathered to watch a steamer come into the harbor. The news of a steamer heading into Michigan City on such a stormy day spread like wildfire, as it probably meant she was in trouble. Hundreds of locals waited on the beach to watch the drama. On this snowy, blustery day, the old sailor gazed out into Lake Michigan, saw the tall thin line of smoke pouring from the ship's smokestacks and shook his head. A burst of snow suddenly blocked his view of the steamer as the waves of the lake pounded the breakwater. The sailor commented out loud, "She must be in trouble." It was a general statement directed to anyone who would listen to him. The old skipper thought that no one would be crazy enough to attempt to bring a ship of that size into this harbor unless he was in serious trouble. Pulling up his coat collar to protect the back of his head from the icy blast coming off the lake, he headed back home. He did not want to watch what he was sure was about to happen.

On this day, October 26, 1898, the ship everyone watched was the *Horace A. Tuttle*. The *Tuttle*, officially No. 95908, was a big wooden steamer built in Cleveland, Ohio, in 1887 by the Cleveland Dry Dock Company. She was originally built for H.J. Johnson. At 250 feet and with 38.8-foot beam, she was propelled through the water by a 2-cylinder steam engine. When fully loaded with 1,355 net tons of cargo, she would draw a depth of 20 feet. Like

The *Horace A. Tuttle. Michigan City Public Library Collection.*

many ships that sail the Great Lakes, she was bought and sold by several owners; her last owner was the Nicholas Transit Company. When launched with her three masts and two smokestacks, she was considered the height of shipbuilding technology and an object of considerable beauty for a cargo-carrying steamship.

That fateful trip started in Chicago, Illinois, as she loaded her hold with corn. Corn is one of the cargos that is hard to transport for two reasons. First, if the ship is in a bad storm with high waves, the cargo can shift or move in such a way that it throws the ship into a list. This can cause the ship to be exposed and allows the waves to climb aboard easily. Second, corn is a problematic cargo because when it gets wet, it expands, exerting high pressure on the hatches and hull. There are cases when a cargo hatch has blown completely off a ship or the sides of a ship have split open with the force of dynamite going off in the ship's hold. Only ships that are known to have a dry hold were allowed to take corn as cargo. However, the *Tuttle* was only eleven years old and in her prime. Her hatches and equipment were in very good shape. The crew on the *Tuttle* for this trip included the following:

Tuttle's Crew

Name	Rank	Hometown
J.C. Thompson	Captain	Cleveland
W.J. Fitzgerald	First Mate	Marine City
Gilbert King	Second Mate	Cleveland
E.B. Butler	Chief Engineer	Cleveland
R.O. Butler	Second Engineer	Cleveland
H.T. Foster	Fireman	Cleveland
James Burton	Fireman	Cleveland
John McIntyre	Wheelsman	Cleveland
Walter McDougall	Wheelsman	Superior
Fred Cooley	Watchman	Port Huron
James Brady	Sailor	Cleveland
John Starkey	Sailor	Cleveland
William Snyder	Cook	Cleveland
Mrs. Wm. Snyder	Assistant Cook	Cleveland

A superstitious sailor might predict that the *Tuttle* was doomed because she had a woman on board. However, this was not a prevalent belief on Great Lakes vessels. On the contrary, crews usually welcomed women working as cooks, assistant cooks and stewards. They typically cooked better than their male counterparts. Additionally, if they were working with their husbands, it added a little more income to the family's wages.

As the *Tuttle* left Chicago, the barge *Aberdeen* was tied to her stern. Most barges of this era were old, worn-out, wind-grabbing schooners. They were generally pushed past their usefulness by having their masts cut down and seeing the end of their lifetime attached to the end of a towline. The *Aberdeen* was different: she was built for the express purpose of being towed by a steamer. She was built in Bay City, Michigan, in 1892 by the famous James Davidson. His shipbuilding yard was well known around the lakes. He built big wooden ships made from the oak close to his shipyard. Every time Davidson put a steamship or barge into the water, his critics would say that it was the biggest one he could build. Then, usually, his next ship or barge would prove them wrong. When the *Aberdeen* hit the water, she was named *Gladys H.* (registered no. 106,975) and had a length of 211 feet. She had a beam of 35 feet, was able to carry a net cargo of 993 tons and drew a depth of only 16.6 feet. The *Gladys H.* was initially owned by Davidson Shipping Company, as was a custom of James Davidson. If he didn't have an immediate buyer of the ship he was building when launched, he would take her into his own fleet and then sell her at a future date. Over the years, she was owned by many different companies, and in 1898, she was purchased by the Minch Transportation Company of Cleveland, Ohio.

On Monday, October 24, the wind blew lightly from one direction and then another under sullen gray clouds. The *Tuttle* and her tow, *Aberdeen*, left for Chicago to deliver their cargo in Buffalo, New York. Heading north in Lake Michigan, the crew settled into the routines of their jobs. The aroma of freshly baked bread wafted from the galley, making everyone in the aft end of the *Tuttle* hungry. The cook, Bill Snyder, and his wife started to get the mess ready for dinner. With the waves beginning to grow, the task became increasingly difficult. After dinner was served and the galley cleaned up, Cook Snyder was thankful that he was able to get everything stowed away because the weather was starting to turn into a full-blown gale. Looking back at the *Aberdeen*, he was thankful that he was on the *Tuttle*. He saw how she was pounding in the waves at the end of the towline, and he felt sorry for

The barge *Aberdeen*. *Alpena County George N. Fletcher Public Library, Great Lakes Maritime Collection.*

the crew experiencing a rough ride. About 11:00 pm., Captain Thompson noted in the log that they were in a full gale, and the wind was coming out of the northwest. With the waves running high and the ship pounding, they had had reasonably good weather most of the day on Tuesday. Then, just off of Big Sable Point, north of Ludington, Michigan, Chief Engineer Butler made the perilous trip from the aft-end engine room to the pilothouse. What he had to say to the captain was worth risking his life. He told Captain Thompson that the ship had sprung a leak, and the water was rising in the hold. He had the pumps running and a man looking for the source of incoming water, but he was having trouble keeping the water level down.

Captain Thompson decided that he had to continue on his course, even though he wanted to turn around and head back to Chicago for repairs. But with the current mood of Lake Michigan, turning around would be suicidal for the *Tuttle* and the *Aberdeen*. Turning would require them to be in the wave troughs for a considerable amount of time. That would expose the *Tuttle* and *Aberdeen* to angry waves sweeping over their sides. Additionally, with the *Tuttle*'s limited engine power, there was a distinct possibility that she would not be able to pull both her and the *Aberdeen* out of the wave troughs. There was the added danger that if the *Tuttle* were able to head again into the waves pulling the *Aberdeen* out, it would exert tremendous strain on the towline. If that snapped, the *Aberdeen* would be on her own. With the current conditions on the lake, it would be impossible for them to reconnect the line.

So, he continued on a northerly course, and around midnight, the lake decided for him. Unable to hold the *Tuttle* into the wind, it snapped her into a wave trough. After about fifteen minutes of struggling, with the big barge tied to her, the captain ordered his crew to take an axe to the towline before they were completely swamped. Nowadays, it sounds heartless for a steamer to cut loose a barge in a storm. But in most cases, the barge usually had a better chance of survival. The steamer *L.R. Doty* cutting loose the *Olive Jeanette* in a storm on Lake Michigan is a good example. Although the *Olive Jeanette* survived the storm, she was battered all night and was found just north of Chicago. Towed into the port, she discovered that the *L.R. Doty* had disappeared and probably sank. Another instance of a steamer cutting loose a barge was that of the steamer *Mataafa* in Lake Superior. She cut loose her barge, the *James Nasmyth*, right before she tried to enter the Duluth Harbor. The *James Nasmyth* was able to drop anchor and ride out one of the worst storms on Lake Superior, while the *Mataafa* hit one of the piers trying to enter the harbor. Sinking just north of the piers in Duluth, nine men in the aft end of the ship froze to death.

After cutting the towline, the *Aberdeen* briefly could be seen and seemed to be riding well, even though she was pitching in the waves considerably. The *Tuttle* headed for the west shore of Lake Michigan and started all seven of her pumps. She was leaking badly, but the seven pumps held the rising water in her hold in check. They were hoping to make it to Milwaukee and shelter there.

The situation changed considerably around 5:00 a.m. on Wednesday. A colossal wave smacked the *Tuttle* on her side, taking several of the hatches, her deckhouse and one of her yawl boats over the side. The deckhouse was in the center of the steamer on her main deck. With hatches missing, any water coming on deck would have free access to her hold and drench the corn in an enclosed space. The captain determined that the *Tuttle* was closer to Michigan City than any port on the west coast of Lake Michigan. As the waves rolled over her decks, the crew labored with the determination of those facing death. They tried to hinder the water flow through the hatches into her cargo hold with anything loose. The crew tore off the canvas used for sails on her masts, along with mattresses, blankets, quilts and boards, but nothing was sufficient to stem the water flow. Next, the men, in pure desperation, broke out the hand pumps and got them going. They knew that it was not a question of keeping the *Tuttle* afloat; it was a question of how long they could keep her from sinking to the bottom. The only thing they could do now was slow the incoming water long enough for them to get in or near a safe harbor. Captain Thompson knew that he was close to the Michigan City Harbor, but he also knew that he was drawing too much depth to make through the piers.

When Captain Thompson left the port of Chicago, the *Tuttle* was drawing fourteen feet of water at her bow and sixteen feet at her stern. With about six feet of water in her holds, she was drawing sixteen feet at her bow and more than twenty feet at her stern. He hoped to get close enough to the harbor that some of the *Tuttle* would remain above water so the brave lifesavers there could rescue them. It was now a countdown to how long it would be before the water got high enough to put out the fires below her boilers, thus ending any hope of propulsion and keeping her afloat. He knew that at the rate the water was rising in her holds, he had about three hours to get close to Michigan City or she would sink in deep water. Captain Thompson worked hard to save his boat; he hadn't left the wheelhouse, except for short moments to help his crew, since about 3:00 p.m. on Tuesday. Twenty-four hours later, Captain Thompson saw that they were getting close to the entrance at Michigan City. But he had only a slight hope that the waves would carry them over the sandbars near the entrance of the harbor.

The momentum of slamming into the first sandbar spun them around with the stern facing the shore. Loud reports echoed through the air as the rudder and rudder stock snapped off like dry twigs when they impacted the hard sand bottom of Lake Michigan. Chief Engineer Butler was in the engine room when she struck. The impact was so hard that it knocked him prostrate on the deck. Also, one of the steam pipes burst, filling the room with scalding hot steam. As the second engineer rushed to his side, he told him that they needed to get out of that hellhole before the steam burned their lungs. Helped by the second engineer and one of the two firemen, they crawled out on their hands and knees. Once outside, the *Tuttle* was in the process of spinning around. But they were close to the Michigan City pier and made a desperate jump for safety. With the energy of desperation, both engineers and the two firemen timed their jump with the waves. Landing on top of the pier, they watched their ship as it slammed down the unmovable breakwater. Helpless and broken, the *Tuttle* finally settled on the bottom with her stern near the breakwater. Her bow was just inside the harbor entrance, about twenty feet away from the pier. She was broken in two and hogged in the middle from the force of the swollen corn, and it was now the job of the local lifesavers to rescue the crew.

Theodore Hemson, the lifesaver in the station's tower, had been watching the *Tuttle* approach ever since the long trail of smoke loomed over the horizon. As the *Tuttle* came closer, he hung a red flag in the tower to tell the steamer's captain that the *Tuttle*'s present course was dangerous. Hemson could see that the *Tuttle* was sitting very low in the water and would not be able to make the harbor entrance. Either Captain Thompson did not see the signal because of the falling snow or he had other problems. Either way, he kept the *Tuttle* on a course for the mouth of the harbor.

Before the *Tuttle* slammed into the first sandbar, Theodore Hemson rang the alarm, which alerted the lifesavers to action. They included the following men:

U.S. Lifesaving Crew, Michigan City

Name	*Rank*
Allen Kent	Keeper
Allen Hullings	No. 1 Surfman
Thomas Armstrong	No. 2 Surfman
Edward Kuhs	No. 3 Surfman
Charles Allen	No. 4 Surfman
Hendrick Heisman	No. 5 Surfman

John Sammet	No. 6 Surfman
Frank Partridge	No. 7 Surfman
Theodore Hemson	No. 8 Surfman

Captain Kent was not idle when he heard the first reports about the *Tuttle* approaching Michigan City; he had gone up into the tower and evaluated the situation. He could tell that the big steamer was in trouble when he saw how low she was settling in the water. He ordered his men to get the equipment ready for a lifeboat rescue without launching the boat. When Hemson finally rang the alarm, it was just a matter of launching it. Once hitting the water, Captain Kent knew that it was a matter of life and death. The lifesavers pulled on the oars with all their energy and headed out through the channel. Approaching the *Tuttle* from bow on, Captain Kent maneuvered the large lifeboat to the lee side of the steamer. As they pulled close, one man jumped from the steamer into the lifeboat. The momentum of the jump pushed the lifeboat away. Captain Kent brought the lifeboat next to the steamer again, and another person jumped and successfully landed in the boat. Again, the momentum of the person landing in the boat pushed it away from the steamer. Once again, Captain Kent had his lifesavers pull the lifeboat up to

Michigan City's lifesaving crew. *Michigan City Public Library Collection.*

Michigan City's lifesaving station. *Michigan City Public Library Collection.*

the steamer. This effort was repeated nine times until only Captain Thompson and the crew's black bulldog, Prince, were left on the deck. Prince had sensed the trouble they were in and had been howling the entire time, fearing that he would be left to perish alone. It was a dismal scene—cloud-covered sky and a heavy snow falling. Hearing the dog's howls, the sounds of the wind and waves crashing, Captain Thompson picked up the frightened animal and tossed him into the lifeboat. After exhausting maneuvering, Captain Thompson heard the lifesaving crew over the screaming wind, telling him to jump. He did and landed safely in the lifeboat. For the first time in more than a day, the captain was relieved of the weight of imminent death. And equally important, his crew was finally safe.

Back at the station, the lifesavers and the *Tuttle*'s crew gather around the hearth, where a roaring fire warmed and dried everyone. They all laughed as Prince and the station's dog scampered and played together. The four men who had jumped from the *Tuttle*'s stern onto the Michigan City pier now joined their shipmates in a joyous reunion, each with a story of survival to tell.

Once warm and dry, Captain Thompson approached Captain Kent and asked if some of the *Tuttle*'s crew and the lifesavers would consider going back out to the wreck. When the crew had abandoned the *Tuttle*, they left their personal belongings on board and now wanted to retrieve them. Relaunching the lifeboat, Captain Kent and his crew rowed Captain Thompson and several of the *Tuttle*'s crew back to the wreck. They were able to gain the deck on the forward part near the pilothouse. On board, Captain Thompson's men retrieved the baggage from the crew's quarters in the bow while he got the ship's papers and logs. It was too dangerous to go to the quarters in the stern. So, the four men who came from the engine room and Mr. and Mrs. Snyder would just have to be content with being alive and losing only their personal belongings. As they rowed back to the station, everyone—the lifesavers and the *Tuttle*'s crew—was exhausted. Once there, Captain Kent learned that his number one surfman had already arranged rooms at the Brinkman Hotel for the *Tuttle*'s crew. Once the equipment was back at the lifesaving station, both crews decided to call it an early evening and get some sleep.

The next month, Summer Kimball, the head of the U.S. Life-Saving Service, received the following letter from Captain Thompson:

> *Cleveland, Ohio, November 7, 1898*
>
> *Dear Sir,*
> *I am the late master of the Horace A. Tuttle, which was wrecked at Michigan City, Lake Michigan, on October 26, 1898, after a long battle with one of the worst storms I have ever experienced on the lakes or anywhere, except off Cape Horn and Good Hope. I finally had to run for the above-named port to try to save the lives of the crew, hoping the steamer would keep afloat until we reached that place, which she did, but if it had not been for the able manner in which the lifeboat was handled, by Captain Kent and his crew, we would have perished, as the seas were mountain high at the time.*
>
> *Through you, I take the opportunity to thank them for the skillful manner in the way they took us off the Tuttle.*
>
> *I am very truly yours,*
> *John C. Thompson*
> *Master of Steamer Horace A. Tuttle*

So what happened to the *Aberdeen*? She had run aground near Grand Haven, Michigan, and her crew was rescued by the lifesavers there. On November 8, 1898, she was pulled off the beach by the steamer *Nyack*

The *Horace A. Tuttle* wreck. *Michigan City Public Library Collection.*

and then towed over to Milwaukee to unload her cargo of barley and be placed into dry dock. She sailed the Great Lakes under several owners until 1930, when her last owners, Consolidated Oka Sand & Gravel, closed her registration and abandoned her.

The Nicholas Transit Company abandoned the *Tuttle* where she sat. Since she was only eleven years old, she was insured for $72,000. The Armour & Company of Buffalo, New York, owned her seventy-seven thousand bushels of corn and had insured them for $27,000. Captain James Reid of Reid Wrecking Company won the bid to remove the wreck after the government declared that the *Tuttle* was a hazard to navigation because it partially blocked the entrance to Michigan City Harbor. On May 5, 1899, Captain Reid in the tug *Salvor* removed the last piece of the *Tuttle*, took it out to deeper water and let it sink. The U.S. government paid the Reid Wrecking Company $2,900 to remove the wreck.

A Little More Time...

December is always a bad month for sailing in the Great Lakes. Ice, cold, snow, fiercely blowing winds and enormous waves all create a terrible environment. Captain George Trotter of the bulk carrier *F.W. Wheeler* was outside the pilothouse in Buffalo, New York, giving orders to cast off her lines for a trip to Chicago, Illinois. Leaving the Lake Erie port on Wednesday, November 30, 1893, he would have to sail the entire length of Lake Erie, up through Lake St. Clair at Detroit and then around the east side of the Michigan mitten—the whole length of Lake Huron. At the top of the mitten, Captain Trotter would go through the Straits of Mackinac and enter the top of Lake Michigan. By then, it would be December. How he hated making these trips at this time of year, but Chicago needed the 2,100 tons of hard coal in her hold to warm the homes there and keep businesses running through the winter. He would be glad when this long trip was over.

The *F.W. Wheeler* was built by the eponymous company, F.W. Wheeler & Company, and was launched in 1887 in West Bay City, Michigan. David Whitney Jr., the original owner, had the Wheeler Company build her to a length of 265.5 feet and a beam of 40.5 feet. She was considered big for a wooden ship and was powered by the latest triple-expansion steam engine, which produced eight hundred horsepower. This was at a time when sailboats were converted to steam propulsion, but she still had four masts mounted on her hull. This way, she could raise sail when the winds were favorable to help her conserve fuel. It sounded like a great idea but was rarely practiced. The *F.W. Wheeler* was built to take 1,175 tons of cargo into her holds.

Leaving Buffalo, Captain Trotter rang up the *Wheeler*'s standard speed of 12 miles per hour. He headed west on a course that would bring him to the mouth of the Detroit River and through Lake St. Clair. Taking about nineteen hours to run the length of Lake Erie, he expected to pass through Lake St. Clair early on Thursday morning. After entering the river, he would have to reduce his speed to navigate the 86 miles up that river/lake, entering Lake Huron at the port city of Port Huron, Michigan. From there, he would navigate north-northwest for 167 miles before turning on a more westerly course to go through the Straits of Mackinac and enter Lake Michigan. Ship captains strongly disliked Lake Erie because her shallow waters are easily agitated in almost any amount of wind, and in late November and early December, the lake is never in a good mood. Captain Trotter would be happy to get into the deeper waters of Lake Huron and hug the west coast to give him some protection from the wind. After about twenty-nine hours of sailing, he made it into Lake Huron through wind, waves and snow. Watching the falling barometer on the pilothouse wall, he knew that a storm was coming and hoped that it would hit after he was in Chicago.

After almost another three hundred miles from Port Huron, he put the *Wheeler* on a southerly course. She had handled the bad weather all through the trip. At only six years old, she was considered a new ship and had minimal leakage and tightly clamped hatches. Experiencing no trouble so far, the captain was proud of the way she shook off the waves. Even though Captain Trotter had sailed this course countless times since he was a master, he still did not like doing so in bad weather or so late in the sailing season. Making his turn south into Lake Michigan, snow started falling again.

By Saturday, December 2, around 8:00 p.m., the captain calculated that he was about ten miles off Waukegan, Wisconsin, even though he could not see any landmarks. By that time, the snow was coming down heavily, and the seas were running high.

Today, we are so used to technological conveniences at our fingertips—with a press of a button, we can know our exact position on a map. But in the late 1800s, the most advanced technology Captain Trotter had was a clock. Guessing his speed and compass bearing from his last turn, he calculated that he had traveled 252 miles in the previous twenty-one hours. Based on the current conditions of nearly whiteout snow and rough waves, he realized that at the current speed, there was no way he could safely travel the remaining 35 miles to Chicago. So, he reduced the *Wheeler*'s speed to a point where she was manageable but making very little headway. He also decided to turn the *Wheeler* on more of an eastern course away from the

The *F.W. Wheeler*. *Alpena County George N. Fletcher Public Library, Great Lakes Maritime Collection.*

west coast of Lake Michigan. This route allowed the *Wheeler* to run with the waves and the wind, making it a smoother ride. He was stalling for time when the sea, weather and visibility conditions would improve enough that he could safely enter Chicago Harbor.

Unfortunately, Captain Trotter did not consider the winds pushing the *Wheeler* farther south and east than his estimated position. Knowing that his position was probably off, Captain Trotter ordered depth soundings at regular intervals. Most of the time, he had around 240 feet of water underneath him. In this part of Lake Michigan, the farther you go to the southeast, the shallower the bottom slowly gets. Then, when you are only about one thousand yards off the beach, the bottom quickly goes from 100 to 50 feet of water to less than 20 feet. Between the time when fifteen fathoms (90 feet) sounded and the time of the next sounding, it changed to three fathoms (around 20 feet). Again, it is difficult for us to imagine how this could have happened since we now instantaneously access information about our position and depth with electronics.

Captain Trotter had to rely on a knotted and weighted line. The line was thrown over the side, and the knots were counted until the weight hits bottom. The line was retrieved, and the whole process started again. All this took time, and even though the *Wheeler* was traveling slowly, she still moved a lot of distance before the weighted line hit the water again. When Captain Trotter heard they were in three fathoms of water, he felt the first bump of the bow hitting the hard sand bottom. He ordered the helmsman to turn and called to the engine room for more speed on the propeller. As the ship swung around, he felt the sickening *thud* of the stern hitting bottom, and the *Wheeler* bumped toward the beach. As if to mock the *Wheeler*'s crew, the veil of snow that had blinded them suddenly lifted. They could see that they were only a few hundred feet away from shore. At the same time, the lookout in the Michigan City Life-Saving tower fired a flare high in the air and out to sea by to warn him of the danger. Unfortunately, it was too late to do anything because he was solidly grounded in the shallows. As the *Wheeler* bumped farther toward shore, Captain Trotter shook his head and wished that he had seen that signal earlier. Then he would have had a little more time to prevent the *Wheeler* from wrecking.

It started snowing heavily again, which blotted out all visual references around the *Wheeler*. Captain Trotter was determined to get her afloat again and ordered his crew to start lightening her. They opened her hatches and started to throw her coal cargo over the side. After about one hundred tons of hard coal went over her side, Captain Trotter managed to lighten the *Wheeler* enough to get her off the bottom and into deeper water. At first, it appeared that his efforts were successful, but then the *Wheeler* hit an outer sandbar and got stuck again. From her movements, he could tell that the *Wheeler* was balanced on the sandbar with the bow and stern hanging free. He tried swinging the bow back and forth to get her off the sandbar, acting as a pivot point in her center. Having swung her around to the east with the waves hitting her broadside and coming aboard, he could not move her anymore. Taking soundings of the depth of water, he found he had only sixteen feet of water at her bow and stern and fourteen feet at her midsection. In his final act of desperation, he ordered his chief engineer to open her seacocks and flood her, so the *Wheeler* wouldn't pound on the bottom and break up. This allowed the water to put out the fires in her boilers. The ship was without heat in any of her rooms, making for a miserable night for the sixteen crewmen. All they could do now was to huddle together and wait for rescue.

Captain Henry Finch, the first keeper of the Michigan City Life-Savers, was selected to organize and train a crew in April 1889. He left the service in April 1897 when his number one surfman, Allen Kent, became the keeper. Michigan City, roughly thirty-two miles away by water on the southern end of Lake Michigan, tended to get a lot of business from ships trying to enter the port of Chicago and the fledgling steel mills in Gary, only twenty-four miles away.

Captain Finch was very busy the night that the *Wheeler* had gone aground. The beach opposite of the *Wheeler* was about three miles from the station. Captain Finch wanted to get his surfboat to that area to launch and perform a rescue. With hired teams of men and horse-drawn snowplows, it took hours to clear a road to the spot opposite the *Wheeler*. Still, it took a lot of effort to haul the surfboat the three miles through a winter snowstorm that at times saw whiteout conditions. Once there, the surfboat had to be lifted on top of a wall of ice more than ten feet high. Captain Finch launched the lifeboat about midmorning, and they were successful getting to the lee side of the *Wheeler*. As they pulled up, instead of rescuing the crew, Captain Trotter asked if they could telegraph Chicago for tugs and pumps. Captain Trotter felt that with the help of tugs, the *Wheeler* could be pulled off the sandbar and towed into Chicago for repairs. The lifesavers made it back to shore, lifting the heavy surfboat back atop the wall of ice on the beach. Captain Finch dispatched one of his men back to Michigan City to telegraph the *Wheeler*'s owners for them to send tugs, as requested by Captain Trotter. The

The *F.W. Wheeler* broken on the beach. *Alpena County George N. Fletcher Public Library, Great Lakes Maritime Collection.*

lifesavers on the beach looked around and found enough wood to start a fire. At least they would be warmer than those on the *Wheeler*.

Around noon on December 3, the winds increased in velocity again, and they heard several cracking sounds coming from the *Wheeler*. As they looked out into the lake through limited visibility, they saw a distress signal go up on the *Wheeler*. The lifesavers again climbed the ten-foot wall of ice and launched the surfboat. As they pulled closer to the *Wheeler*, they could tell that she was starting to break up. With a thunderous report, they watched the stern mast snap off and go over the side. As they got nearer, they could tell that her midsection was bent upward, and a large crack had formed in her hull about amidships. The combination of waves and being suspended in the middle put too much stress on the hull, and it snapped like a twig. Captain Finch only allowed eight of her sixteen-man crew to board the lifeboat. They would have to make two trips in that confused and dangerous sea to rescue all the *Wheeler*'s crew. Getting the men on top of the wall of ice was challenging and exhausting. Launching the boat, once again, they rowed out to the *Wheeler*. Pulling up to the lee side for the third time that day, they rescued the last remaining seven crewmen and Captain Trotter, who was the last man to leave the *Wheeler*. He first made sure that all his men were accounted for and safely in the lifeboat. After getting the remaining crew over the top of the ice wall, they followed the first group and gathered around the roaring fire. The lifesavers pulled their lifeboat over the wall and dragged it a safe distance onto the beach. Figuring that his men would be requested to help once the tugs got to Michigan City, Captain Finch elected to leave the lifeboat on the beach. His men and the crew of the *Wheeler* made the three-mile journey back to the station and its warmth and protection. As the two crews made their way through the snow accumulating already on the plowed road, Captain Trotter told Captain Finch how grateful he was for the lifesavers. But he believed that with a little more time, he may have been able to save the *Wheeler*.

The following evening, the Lenderson House in Michigan City hosted a banquet for Captain Finch and his brave crew. Captain Trotter and the crew off the *Wheeler* hosted the gathering to show their appreciation for the risks this crew undertook to perform this noble rescue. First Mate James Buchanan of the *Wheeler* was the primary planner of the banquet, but all the crew men of the *Wheeler* were enthusiastic in helping with the arrangements. At 9:00 p.m., a bountiful dinner was laid out in the hotel's dining room, and thirty people sat down to eat. Captain Trotter sat at the head of the table, with Captain Finch on his left. Captain Tricket of the steamer *D.C. Whitney* was elected to be the toastmaster. After a joyous dinner and many toasts to both crews, a hat was passed for a collection for the meal's servers for their excellent service. Captain Trotter gave the final toast of the evening to Captain Finch and his courageous men, telling them that they will never be forgotten in the hearts of the men of the *Wheeler*.

That Sunday evening, after several days with the men off the tugs and the lifesavers having lightened the *Wheeler* to pull off the sandbar, it was decided to abandon any further work on her. They informed the owners and underwriters of their decision. Because it was so late in the season, only one of the owners, D.J. Whitney, had covered his share in the boat with $37,000 insurance. The other owners had not invested in insurance for this trip, so they completely lost their investment in the *F.W. Wheeler*.

The *Wheeler* sat where she was abandoned all winter in 1893, and the following spring, the tugs returned from Chicago to salvage the engine, boilers, some of her coal cargo and anything else that was of value not lost to the winter storms. After they left her, she slowly broke up from the waves and ice in Lake Michigan.

The Old Man

The old man stood alone on the lakeshore, looking at the million flashes of light reflecting off the dancing waves at sunset. He stood there watching the sky change from calming blue to a brilliant heart-pumping red as the sun dipped below the endless freshwater sea. The lake quenched the flaming fireball in a dazzling display of reds, yellows, oranges and blues. For most people, seeing this beautiful sunset would make their hearts dance. But to the old man, it brought a lonely, empty feeling back to his aching heart.

That emptiness used to be filled with the love of his life. He remembered all the years he spent coming to this beach with her to watch the sunsets and how wonderful her hand felt in his hand. It was so magnificent to stand there talking about all their dreams and hopes. They discussed splendid places to see together and all the beautiful times they had already spent together. In those moments, the future was an exciting and hopeful time. Each day was always better than the day before. It was a glorious feeling to know that she caused the beating of his heart. Just the sheer excitement of her standing next to him and sharing yet another special moment made his heart race to levels he never dreamed possible.

He remembered the time before she came into his life, the cynical years of his youth when he thought that his interest in a woman could not last more than a few months. He remembered telling himself that this thing called love that everyone talked about could only be real for a short time. He never thought it was possible to feel that way for years. Then she came into his life, and he forever felt the same excitement and joy of feeling his heart beating as when he first set eyes on her. He kept telling himself that this feeling could not continue, but then it became years and then decades. His cynicism dissolved into an overwhelming joy he could not express in words. Every time he saw her, his heart would dance. The broad smile always on his face was because of her. He knew from the moment they went out the first time what it was like to be completely in love. The sunsets went from an everyday occurrence of his youth to the pure joy she had brought to his heart. He had never heard the sounds of the glorious waves play into his heart and feel the wind dance on his face. All these wonderful feelings were because she was standing next to him and he was holding her hand.

As life together went on, the sound of the waves was replaced by the constant drum of an oxygen machine. And the time watching sunsets was replaced by endless trips to the hospital. Each year, it got worse, and one broken dream piled on top of other broken dreams. His hope that she would recover and walk on the beach with him again was replaced by the hope that she would be with him just one more day. Then one day, he held her cold hand and felt for the first time in his life the intense pain and gaping void in his heart from losing her. He knew that she became an angel, but that didn't stop the pain. Some days, the tears would not stop, as he wished with all his strength that this pain tearing at his insides would end. The only thing that prevented him from ending his pain was the hope that he would eventually join her in eternity. It took endless days, then months and years of tears to finally find the strength to go back to that shore and watch a sunset. He did find some solace as he remembered all the times they stood there holding hands. On his first visit to their special beach, he took a jar of sand and a jar of water from that place. He wanted to share with her a little bit of that place by putting it where she eternally rests—to share with her how much love he still felt in his heart for her. With this small token, he wanted her to know how much he missed her.

A young couple wanted to get to the beach early that morning to pick the best spot to spend the day before the crowds got there. As they walked to the water's edge, they noticed the still, dark shape of the old man. His clothes were cold and wet, and his legs were partially buried in the sand from a night of the waves washing over them. The sheriff's deputies were called, and they immediately decided that there was no foul play. He had died of natural causes. A crumbled rose, partially covered by the sand, was next to him as he lay there on the beach. The young couple noticed a slight smile on his face. And his right hand was gently closed, as if holding someone else's hand. The deputies noted that he died peacefully there on the shore. Perhaps he got to see one more sunset with her before he went to be with her.

Bibliography

What's in a Name?

Annual Report of the Operations of the U.S. Life-Saving Service. Fiscal year ending June 30, 1878, 51–52.

Braesch, Connie, Lieutenant. "Coast Guard Compass." *Coast Guard Heroes*, official blog of the U.S. Coast Guard, October 27, 2010. www.webcitation.org/mainframe.php.

———. "Coast Guard Compass." *Coast Guard Heroes*, official blog of the U.S. Coast Guard, November 8, 2010. www.coastguard.dodlive.mil/2010/11/coast-guard-heroes-joseph-napier.

McNeil, William R. "*Merchant* (schooner), waterlogged, April 27, 1854." Maritime History of the Great Lakes. www.images.maritimehistoryofthegreatlakes.ca/details.asp?ID=39421&n=1.

———. "*Tuscarora* (brig), aground, September 1, 1855." Maritime History of the Great Lakes. www.images.maritimehistoryofthegreatlakes.ca/details.asp?ID=38741.

Michigan Lighthouse Conservancy. "Saint Joseph Life Saving Station." St. Joseph, Michigan, Lake Michigan. michiganlights.com.

Slideserve. "Coast Guard and U.S. Life-Saving Service." December 4, 2013. www.slideserve.com/cora/the-coast-guard-and-u-s-life-saving-service.

U.S. Coast Guard. "Gold Lifesaving Medals." www.history.uscg.mil/Browse-by-Topic/Notable-People/Award-Recipients/Gold-Lifesaving-Medal.

By the Narrowest of Margins

Annual Report of the Operations of the U.S. Life-Saving Service. "Wreck of the *City of Green Bay*." Fiscal year ending June 30, 1888, 20–25.

Rich, Craig. "*City of Green Bay*." Michigan Shipwreck Research Association. michiganshipwrecks.org.
U.S. Life-Saving Organization. "Daily Life." www.uslife-savingservice.org/lifesavers-duties-equipment/daily-station-life.
Wikipedia. "Breeches Buoy." en.wikipedia.org/wiki/Breeches_buoy.
———. "Lyle Gun." en.wikipedia.org/wiki/Lyle_gun.
———. "U.S. Life-Saving Service." en.wikipedia.org/wiki/United_States_Life-Saving_Service.

Same Bloody Storm

Annual Report of the Operations of the U.S. Life-Saving Service. "Awards of Medals." Fiscal year ending June 30, 1889, 53–54.
McNeil, William R. "*Havana* (Schooner), Sunk, October 3, 1887." Maritime History of the Great Lakes. www.images.maritimehistoryofthegreatlakes.ca/51085/data?n=1.
Meyers, Robert C. "*Havanna*, 1887." In *Lost on the Lakes: Shipwreck of Berrien County, Michigan*. Berrien Springs, MI: Andrews University Press, 2003.
West Michigan Luna Library. "South Haven Lighthouse Log." September 29, 1887–October 3, 1887. South Haven Michigan Lighthouse Logs. wmich.edu.
Wikipedia. "Great Lakes Storm 1913." en.wikipedia.org/wiki/Great_Lakes_Storm_of_1913.
———. "*Mataafa* Storm." en.wikipedia.org/wiki/Mataafa_Storm.

One More

Great Lakes Ships. "The *Myosotis*." Alpena County George N. Fletcher Public Library, Great Lakes Maritime Collection. www.greatlakeships.org.
———. "*Myosotis* in Ice." Alpena County George N. Fletcher Public Library, Great Lakes Maritime Collection. www.greatlakeships.org.
Meyers, Robert C. "*Myosotis*, 1887." In *Lost on the Lakes: Shipwreck of Berrien County, Michigan*. Berrien Springs, MI: Andrews University Press, 2003.
Milwaukee County. "Lake Michigan Shipwreck and Disasters: *Myosotis*." www.linkstothepast.com/milwaukee/marineM.php#Myosotis.
Peterson, William. *United States Life-Saving Service in Michigan*. Charleston, SC: Arcadia Publishing. 2000.

Only Grit

Boyne, Jeff. "Armistice Day Storm: November 11, 1940." National Weather Service. www.weather.gov/arx/nov111940.
Rich, Craig. "*Anna C. Minch*." Michigan Shipwreck Research Association. www.michiganshipwrecks.org/shipwrecks-2/shipwreck-categories/shipwrecks-found/anna-c-minch.

———. "*Richard H.* and *Indian.*" Michigan Shipwreck Research Association. www.michiganshipwrecks.org/shipwrecks-2/shipwreck-categories/shipwrecks-lost/richard-h-and-indian.

———. "*William B. Davock.*" Michigan Shipwreck Research Association. www.michiganshipwrecks.org/shipwrecks-2/shipwreck-categories/shipwrecks-found/william-b-davock.

Simmons, Theresa, and Sara Schultz. "Armistice Day Blizzard of 1940 Remembered." National Weather Service. www.weather.gov/dvn/armistice_day_blizzard.

South Haven Daily Tribune. "Body of John Taylor Is Identified." November 18, 1940.

———. "Fear 12 Lost in Lake Michigan: Three Boats Missing; Families Watch from Lake Beach Here." November 12, 1940.

———. "Fear 12 Lost in Lake Michigan: 3 South Haven Boats Missing." November 12, 1940.

———. "*Richard H.* Name Plate Found, 18 Seamen Known Dead, 51 Missing in Storm." November 13, 1940.

———. "Search Continues for Bodies; South Haven Will Welcome Heroic Guardsmen." November 14, 1940.

———. "Searchers Find No Sign of Bodies: 1500 Line Harbor Piers to Cheer Heroic Guardsmen." November 15, 1940.

St. Joseph Herald Press. "Expected Toll in Lake Gale to Reach, at Least 65." November 13, 1940.

———. "Guards Safe: Tell of Icy Rescue." November 13, 1940.

———. "Shore Yields Bits of Lost Fishing Boat." November 13, 1940.

———. "Storm Damages County; Boat Battles Heavy Seas." November 11, 1940.

———. "Summary of Heavy Toll in Shipping." November 13, 1940.

———. "Two Tugs and Coast Guard Boat Missing." November 12, 1940.

———. "Water Yields Torn Bits of S.H. Tugs." November 14, 1940.

Grand Slam in Grand Haven

Annual Report of the Operations of the U.S. Life-Saving Service. "Wreck of the *Australia.*" Fiscal year ending June 30, 1879, 29.

———. "Wreck of the *L.C. Woodhuff.*" Fiscal year ending June 30, 1879, 25–29.

Jackson Citizen Patriot. "Other Wrecks." November 4, 1878.

Regulations of the Life-Saving Service of the United States. "Article V: Drill and Exercise." 1899. Treasury Department no. 2096, Life-Saving Service.

Seibold, David H. *Coast Guard City, U.S.A: A History of the Port of Grand Haven.* Edited by Dorothe Welch Seibold. N.p.: Historical Society of Michigan, 1990.

Shelak, Benjamin J. *Shipwrecks of Lake Michigan.* Saline, MI: McNaughton & Gunn, 2003.

Stonehouse, Frederick. *Wreck Ashore: The United States Life-Saving Service on the Great Lakes.* Duluth, MN: Lake Superior Port Cities Inc., 1994.

Tag, Thomas. "Wreck of the Woodruff by Thomas Tag," United States Lighthouse Society. www.uslhs.org/wreck-woodruff.
Wikipedia. "Summer Increase Kimball." en.wikipedia.org/wiki/Sumner_Increase_Kimball.

Note: A special thank-you to Jeanette Weiden at the Louitt Public Library in Grand Haven for helping me to do the research for this story during the pandemic.

In Search of Jack Dipert's Ghost

Barry, James P. "Wrecks of the Twenties and Thirties." In *Wrecks and Rescues of the Great Lakes: A Photographic History*. N.p.: Howell-North Books, 1981.
Boyer, Dwight. "To Hell with 'Er." In *Strange Adventures of the Great Lakes*. N.p.: Freshwater Press Inc., 1974.
Gaertner, Eric. "75 Years Later, the Sinking of the Steamer *Henry Cort* Remembered." Michigan Live, November 27, 2009. www.mlive.com/news/muskegon/2009/11/75_years_later_the_sinking_of.html.
Great Lakes Ships. "Damaged *Henry Cort*—1927." Alpena County George N. Fletcher Public Library, Great Lakes Maritime Collection. www.greatlakeships.org.
———. "*Henry Cort* Leaving Duluth." Alpena County George N. Fletcher Public Library, Great Lakes Maritime Collection. www.greatlakeships.org.
———. "*Henry Cort* Whirly Cranes." Alpena County George N. Fletcher Public Library, Great Lakes Maritime Collection. www.greatlakeships.org.
———. "*Pillsbury*." Alpena County George N. Fletcher Public Library, Great Lakes Maritime Collection. www.greatlakeships.org.
Lydecker, Ryck. *Pigboat the Story of the Whalebacks*. Superior, WI: Head of the Lakes Maritime Society Ltd., 1981. Reprint, Inland Seas, 1953.
Muskegon Chronicle. "Captain Did All Anyone Could, Crew Says." December 3, 1934.
———. "Freighter Cort Down: Crew Is Believed Dead." December 1, 1934.
———. "Seek Body of Coast Guardsman as Crew of 25 Aboard Cort Rescued." December 1, 1934.
———. "The Sinking of the *Henry Cort* Unforgettable: Events Commemorate 75th Anniversary." November 27, 2009.
Sherman, Elizabeth B. "Breakwater Blues: The Wreck of the *Henry W. Cort*." In *Beyond the Windswept Dunes: The Story of Maritime Muskegon*. Detroit, MI: Wayne State University Press, 2003.
Stonehouse, Frederick. "Hoodoo Ships." In *Haunted Lakes: Great Lakes Ghost Stories, Superstitions and Sea Serpents*. Duluth, MN: Lake Superior Port Cities Inc., 1997.
Van der Linden, Reverend Peter, ed., as well as John H. Bascom, Edward J. Dowling, Reverend Worden, Peter B. Wright and Richard J. Doc. *Great Lakes Ships We Remember*. N.p.: Freshwater Press Inc., 1979.

Not a Good Time

Annual Report of the Operations of the U.S. Life-Saving Service. "Wreck of the *City of Duluth*." Fiscal year ending June 30, 1898, 138–39.
Daily Palladium (Benton Harbor, Michigan). "Cannot Save the Boat." January 29, 1898.
———. "Forty-One Saved." January 29, 1898.
———. "Forty Saved." January 27, 1898.
———. "May Be Saved; *City of Duluth* May Be Released." January 28, 1898.
———. "Wrecked! *City of Duluth* Going to Pieces." January 27, 1898.
Great Lakes Ships. "*City of Duluth* in Harbor." Alpena County George N. Fletcher Public Library, Great Lakes Maritime Collection. www.greatlakeships.org.
Meyers, Robert C. "*City of Duluth*, 1898." In *Lost on the Lakes: Shipwreck of Berrien County, Michigan*. Berrien Springs, MI: Andrews University Press, 2003.
Stonehouse, Frederick. "Manistee." Introduction in *Went Missing*. N.p.: Avery Color Studios, 1984.

Note: A special thank-you to Jeanette Weiden at the Louitt Public Library in Grand Haven for helping me to do the research for this story during the pandemic.

Twice Rescued

Annual Report of the Operations of the U.S. Life-Saving Service. "Wreck of Steamer *Protection*." Fiscal year ending June 30, 1884, 20–27.
Great Lakes Ships. "H.C. *Akeley* Floundering." Alpena County George N. Fletcher Public Library, Great Lakes Maritime Collection. www.greatlakeships.org.
———. "Tug Protection." Alpena County George N. Fletcher Public Library, Great Lakes Maritime Collection. www.greatlakeships.org.
Meyers, Robert C. "*Arab*, *Protection* and *H.C. Akeley*, 1883." In *Lost on the Lakes: Shipwreck of Berrien County, Michigan*. Berrien Springs, MI: Andrews University Press, 2003.
Michigan Research Association. "H.C. *Akeley*." www.michiganshipwrecks.org/shipwrecks-2/shipwreck-categories/shipwrecks-found/h-c-*Akeley*.
Shelak, Benjamin J. *Shipwrecks of Lake Michigan, 1880–1900*. Saline, MI: McNaughton & Gunn. 2003.
VanHeest, V.O. "Convoluted Circumstances." In *Lost & Found*. N.p.: In-Depth Editions, 2012.

Sailors in Peril

Annual Report of the Operations of the U.S. Life-Saving Service. "Letters Acknowledging the Services of the Life-Saving Crews." Fiscal year ending June 30, 1899, 247.
Detroit Free Press. "Sailors in Peril." October 27, 1898.

Evening News (Michigan City, Indiana). "The Loss Is Total." October 27, 1898.
———. "Steamer Abandoned." October 28, 1898.
———."The Work Is Done." May 5, 1899.
Great Lakes Ships. "*Aberdeen* (1892 Schooner-Barge)." Alpena County George N. Fletcher Public Library, Great Lakes Maritime Collection. www.greatlakeships.org.

A Little More Time

Annual Report of the Operations of the U.S. Life-Saving Service. "F.W. *Wheeler*." Fiscal year ending June 30, 1894, 158.
Great Lakes Ships. "*Wheeler F.W.* (1887 Bulk Freighter)." Alpena County George N. Fletcher Public Library, Great Lakes Maritime Collection. www.greatlakeships.org.
———. "*Wheeler F.W.* (1887 Bulk Freighter) Wrecked on the Beach." Alpena County George N. Fletcher Public Library, Great Lakes Maritime Collection. www.greatlakeships.org.
Michigan City (I.N) News. "The Boat Abandoned: The Big Boat Cannot Possibly Be Saved." December 6, 1893.
———. "On the Beach: The Propeller F.W. Wheeler Ashore Near Here." December 6, 1893.

About the Author

Michael Passwater is an electronics engineer with a passion for researching the details of history, especially Great Lakes history. He has spent years researching the history of the U.S. Life-Saving Service and the shipwrecks in the Great Lakes. Researching the intriguing stories of the men and women who played critical roles in the development of the rich maritime history along the east coast of Lake Michigan has led to this publication.